Reframing Special Relativity

Part I

Michael Kunkler

Copyright © 2021 Michael Kunkler
All rights reserved.
ISBN-13: 979-8-72-401973-6

Dedication

For my wife and children

Contents

1. **Introduction** — 1
   - 1.1 Special Relativity — 2
   - 1.2 Midpoint Relativity — 5
   - 1.3 Acknowledgement — 10

2. **Time Intervals** — 11
   - 2.1 Thought Experiment — 11
   - 2.2 Principle of Relativity — 12
   - 2.3 Special Relativity — 13
   - 2.4 Midpoint Relativity — 15
   - 2.5 Relativity Comparison — 23

3. **Length of Objects** — 25
   - 3.1 Thought Experiment — 25
   - 3.2 Principle of Relativity — 26
   - 3.3 Special Relativity — 27
   - 3.4 Midpoint Relativity — 32
   - 3.5 Relativity Comparison — 44

4. **Frequency Shift** — 47
   - 4.1 Thought Experiment — 47
   - 4.2 Principle of Relativity — 48
   - 4.3 Special Relativity — 49

	4.4	Midpoint Relativity	54
	4.5	Relativity Comparison	68

5. Coordinate Transformations   71
	5.1	Special Relativity	71
	5.2	Midpoint Relativity	76
	5.3	Relativity Comparison	86

6. Conclusion   87

Preface

"In the interest of clearness, it appeared to me that I should repeat myself frequently"

Albert Einstein, 1916

This book extends the special theory of relativity by addressing two inherent problems. The first problem is the assumption that one of the two inertial reference frames is in a state of absolute rest. The second problem is the constraint that the relative velocity between two inertial reference frames is less than the speed of light. In the spirit of Einstein, repetition is again frequently used in the interest of clearness.

Michael Kunkler, 2021

1. Introduction

In the **special theory of relativity**, Einstein (1905) showed that time intervals dilate, the length of objects contract and there is a frequency shift in electromagnetic waves. The extent of these changes is dependent on the relative velocity between two inertial reference frames, which are in uniform motion. The increase in time intervals is known as **time dilation**, the reduction in the length of objects is known as **length contraction**, and the frequency shift in electromagnetic waves is known as the **relativistic Doppler effect**. In addition, the relationship between space and time in one inertial reference frame, and space and time in another inertial reference frame is expressed using **Lorentz coordinate transformations**. These transformations show that space and time are not independent, giving rise to the concept of spacetime.

The special theory of relativity is based on assumptions and constraints. The **at-rest assumption** requires that one of the two inertial reference frames is in a state of absolute rest. The question then arises: which inertial reference frame does one choose? The **relative-velocity constraint** restricts the relative velocity between two inertial reference frames to be less than the speed of light. The relative-velocity constraint is not the same as the **light-speed constraint**, which states that objects with mass cannot travel faster than the speed of light. The special theory of relativity is unable to model two inertial reference frames travelling in opposite directions, where each inertial reference frame travels at more than half the speed of light, even though neither inertial reference frame violates the light-speed constraint.

In this book, the special theory of relativity is extended to remove the at-rest assumption for one of the two inertial reference frames and to increase the relative-velocity constraint from less than the speed of light to less than **twice** the speed of light.

1.1 Special Relativity

The special theory of relativity (Einstein, 1905) consists of two postulates.

Principle of relativity
All the laws of physics are the same in all inertial reference frames.

Principle of the invariance of the speed of light
The speed of light in a vacuum is the same in all inertial reference frames and is independent of the motion of the source.

Inertial reference frames move with uniform motion (constant velocity) and do not undergo acceleration. The first postulate asserts that observers in an inertial reference frame cannot know if they are in motion, rather they only know the **relative** uniform motion compared to another inertial reference frame. In addition, there is no concept of a state of absolute rest. Reference frames that undergo acceleration are known as **non-inertial** (or **accelerated**) **reference frames** and are beyond the scope of this book. Thus, throughout the remainder of this book, inertial reference frames are referred to as **reference frames**.

The second postulate stipulates that all observers measure the speed of light independently of any reference frame and the speed of light in a vacuum is a constant. The second postulate asserts that observers in all reference frames will measure the speed of light as the same value, independently of any reference frame's speed. It should be noted that objects with mass cannot travel at, or greater than, the speed of light in a vacuum, which is known as the **light-speed constraint**.

1.1.1 Special at-rest assumption

The special theory of relativity has two **standard** reference frames, each with their own observer. There is an **event reference frame** with an **event observer** and a **relative reference frame** with a **relative observer**. The event reference frame and relative reference frame move with uniform motion. The special theory of relativity does not distinguish between uniform motion in one reference frame and uniform motion in the other reference frame, only the **relative** uniform motion between the reference frames. The speed of the relative uniform motion between one reference frame and another reference frame is known as **relative velocity**. In addition, it is impossible to know which reference frame is at rest. Nevertheless, the special theory of relativity inherently assumes that one of the reference frames is in a state of absolute rest, which is referred to as the **special at-rest assumption**. A result of this assumption is that the special theory of relativity has four different scenarios for the two reference frames.

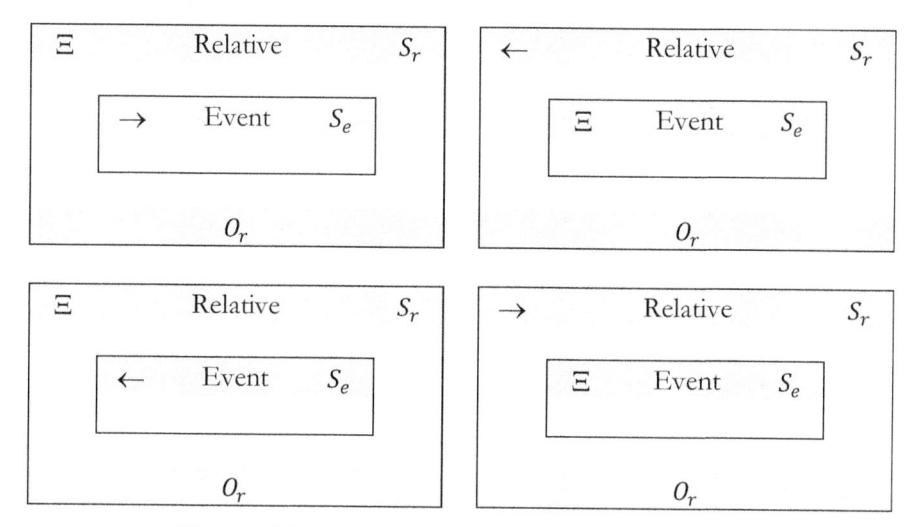

Ξ	Relative	S_r		←	Relative	S_r
	→ Event S_e				Ξ Event S_e	
	O_r				O_r	

Ξ	Relative	S_r		→	Relative	S_r
	← Event S_e				Ξ Event S_e	
	O_r				O_r	

Figure 1-1. The four scenarios of special relativity

Figure 1-1 displays the four scenarios for the two reference frames of special relativity, where S_r represents a relative reference frame, O_r represents a relative observer, S_e represents an event reference frame, Ξ represents a reference frame that is in a state of absolute rest, ← represents a reference frame that is travelling with uniform motion to the left, and → represents a reference frame that is travelling with uniform motion to the right.

The **four scenarios of the special theory of relativity** are:

1. The relative reference frame S_r is in a state of absolute rest Ξ and the event reference frame S_e is in uniform motion to the right →.

2. The relative reference frame S_r is in uniform motion to the left ← and the event reference frame S_e is in a state of absolute rest Ξ.

3. The relative reference frame S_r is in a state of absolute rest Ξ and the event reference frame S_e is in uniform motion to the left ←.

4. The relative reference frame S_r is in uniform motion to the right → and the event reference frame S_e is in a state of absolute rest Ξ.

In each scenario, the special theory of relativity assumes that one of the two reference frames is in a state of absolute rest.

1.1.2 The special relative-velocity constraint

The special theory of relativity constrains the relative velocity between the two reference frames to be less than the speed of light, which is referred to as the **special relative-velocity constraint**. This can be seen by the **special relativistic factor**, which is at the mathematical heart of the theory. The special relativistic factor can be written as:

$$\boxed{\gamma_s = \frac{1}{\sqrt{1 - \frac{v^2}{c^2}}}} \quad (1\text{-}1)$$

where γ_s is the special relativistic factor (or special gamma), v is the relative velocity between the two reference frames, and c is the speed of light in a vacuum ($c = 299{,}792{,}458\ m/s$). The relative velocity can be written as a fraction of the speed of light:

$$v = \beta_s c \quad (1\text{-}2)$$

where v is the relative velocity between the two reference frames, β_s represents the fraction, and c is the speed of light in a vacuum. The special relativistic factor can be rewritten in terms of the fraction β_s by substituting equation (1-2) into equation (1-1) to obtain:

$$\gamma_s = \frac{1}{\sqrt{1 - \frac{(\beta_s c)^2}{c^2}}} = \frac{1}{\sqrt{1 - \beta_s^2}} \quad (1\text{-}3)$$

where γ_s is the special relativistic factor, β_s is the fraction, and c is the speed of light in a vacuum.

When the relative velocity between the two reference frames is equal to the speed of light ($\beta_s = 1$), the special relativistic factor is infinite ($\gamma_s = 1/0 = \infty$). In addition, when the relative velocity between the two reference frames is greater than the speed of light ($\beta_s > 1$), the term in the square root becomes negative, which results in an imaginary special relativistic factor. The special relative-velocity constraint restricts the relative velocity between the two reference frames to be less than the speed of light, and can be written in terms of the fraction β_s as:

$$0 \leq \beta_s < 1 \quad (1\text{-}4)$$

Alternatively, the special relative-velocity constraint can be rewritten in terms of the relative velocity by substituting $\beta_s = v/c$ from equation (1-2) into equation (1-4) to give:

$$\boxed{0 \leq v < c} \qquad (1\text{-}5)$$

where v is the relative velocity between the two reference frames, and c is the speed of light in a vacuum.

However, it is not unreasonable to imagine that two reference frames can travel in opposite directions, where each reference frame travels at more than half the speed of light. In this context, neither of the reference frames violate the light-speed constraint. The special theory of relativity is unable to model this situation ($\beta_s > 1$), and the term in the square root becomes negative, which results in an imaginary special relativistic factor.

1.2 Midpoint Relativity

In a system that consists of two **standard** reference frames in uniform motion, the midpoint is the only point that is in a state of absolute rest. The **midpoint** is defined as the equidistant point between the event being measured in the event reference frame and the relative observer in the relative reference frame.

The special theory of relativity can be reframed by creating a **phantom midpoint reference frame** at the midpoint, together with a **phantom midpoint observer**. The midpoint reference frame is always in a state of absolute rest. Both the event reference frame and the relative reference frame travel with equal speed in opposite directions with respect to the midpoint reference frame.

The creation of a midpoint reference frame divides the relativity environment into two parts: an **event-to-midpoint part** and a **midpoint-to-relative part**. The special theory of relativity can be applied separately to both parts by correctly assuming that the midpoint reference frame is in a state of absolute rest. The solution for each part can be combined to form a complete solution. The reframing of the special theory of relativity is referred to as the **midpoint theory of relativity**.

The midpoint has not gone unnoticed in the special theory of relativity. A thought experiment on simultaneity places observers at the midpoint between two lightning strikes. The synchronisation of clocks can also be achieved with a midpoint clock. Loedel diagrams (symmetric Minkowski diagrams) create symmetry of reference frames around a median (or midpoint) reference frame, so that the two real observers travel with equal speed in opposite directions (Mirimanoff, 1921).

1.2.1 Midpoint at-rest assumption

The midpoint theory of relativity creates a phantom midpoint reference frame with a phantom midpoint observer. This results in three reference frames: two **standard** reference frames and one **phantom** reference frame, where each reference frame has an observer. There is an **event reference frame** (standard) with an **event observer**, a **midpoint reference frame** (phantom) with a **midpoint observer**, and a **relative reference frame** (standard) with a **relative observer**. The midpoint reference frame is the only reference frame that is in a state of absolute rest, which is referred to as the **midpoint at-rest assumption**. A result of this assumption is that the midpoint theory of relativity has two different scenarios for the three reference frames.

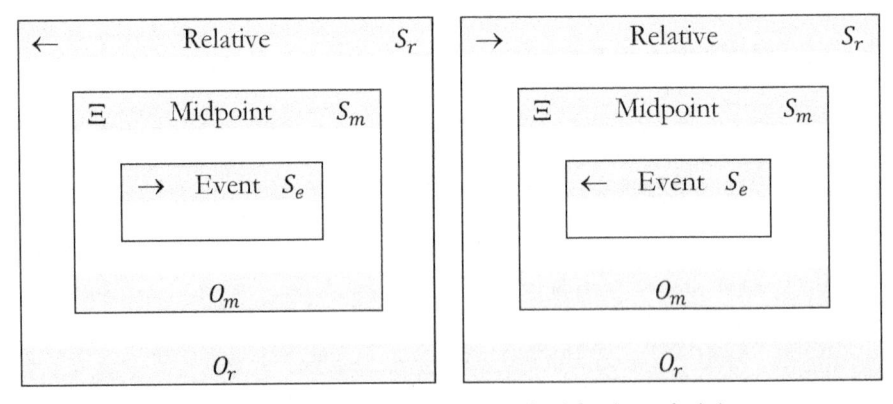

Figure 1-2. The two scenarios of midpoint relativity

Figure 1-2 displays the two scenarios for the three reference frames of midpoint relativity, where S_r represents a relative reference frame, O_r represents a relative observer, S_m represents a midpoint reference frame, O_m represents a midpoint observer, S_e represents an event reference frame, Ξ represents a reference frame that is in a state of absolute rest, $\leftarrow$ represents a reference frame that is travelling with uniform motion to the left, and $\rightarrow$ represents a reference frame that is travelling with uniform motion to the right.

The **two scenarios of the midpoint theory of relativity** are:

1. The midpoint reference frame S_m is in a state of absolute rest Ξ, the event reference frame S_e is in uniform motion to the right $\rightarrow$ and the relative reference frame S_r is in uniform motion to the left $\leftarrow$.

2. The midpoint reference frame S_m is in a state of absolute rest Ξ, the event reference frame S_e is in uniform motion to the left ← and the relative reference frame S_r is in uniform motion to the right →.

In each scenario, the midpoint theory of relativity assumes that only the midpoint reference frame is in a state of absolute rest.

The creation of a midpoint reference frame divides the relativity environment into two parts: an **event-to-midpoint part** and a **midpoint-to-relative part**. The event-to-midpoint part assumes that the midpoint reference frame is in a state of absolute rest and the event reference frame is travelling with uniform motion either to the left or to the right.

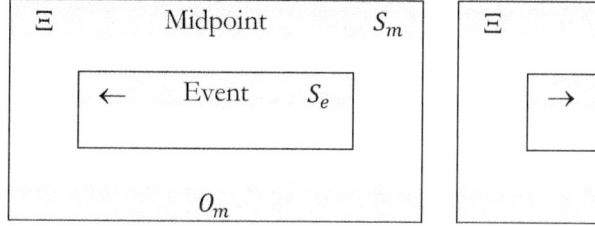

Figure 1-3. The two scenarios of the event-to-midpoint part

Figure 1-3 displays the two scenarios for the two reference frames of the **event-to-midpoint part** of midpoint relativity, where S_m represents a midpoint reference frame, O_m represents a midpoint observer, S_e represents an event reference frame, Ξ represents a reference frame that is in a state of absolute rest, ← represents a reference frame that is travelling with uniform motion to the left, and → represents a reference frame that is travelling with uniform motion to the right.

The **two scenarios of the event-to-midpoint part** of the midpoint theory of relativity are:

1. The midpoint reference frame S_m is in a state of absolute rest Ξ and the event reference frame S_e is in uniform motion to the left ←.

2. The midpoint reference frame S_m is in a state of absolute rest Ξ and the event reference frame S_e is in uniform motion to the right →.

The midpoint-to-relative part assumes that the midpoint reference frame is in a state of absolute rest and the relative reference frame is travelling with uniform motion either to the left or to the right.

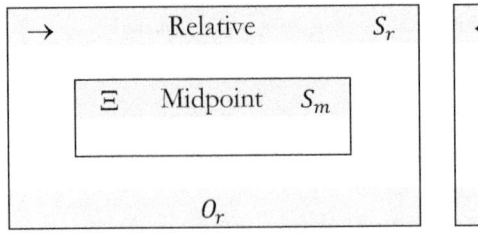

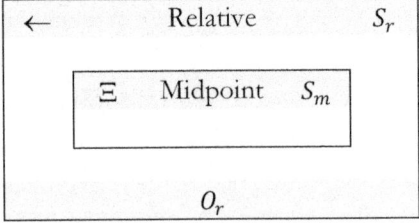

Figure 1-4. The two scenarios of the midpoint-to-relative part

Figure 1-4 displays the two scenarios for the two reference frames of the **midpoint-to-relative part** of midpoint relativity, where S_r represents a relative reference frame, O_r represents a relative observer, S_m represents a midpoint reference frame, Ξ represents a reference frame that is in a state of absolute rest, ← represents a reference frame that is travelling with uniform motion to the left, and → represents a reference frame that is travelling with uniform motion to the right.

The **two scenarios of the midpoint-to-relative part** of the midpoint theory of relativity are:

1. The midpoint reference frame S_m is in a state of absolute rest Ξ and the relative reference frame S_r is in uniform motion to the right →.

2. The midpoint reference frame S_m is in a state of absolute rest Ξ and the relative reference frame S_r is in uniform motion to the left ←.

The special theory of relativity can be applied separately to both the event-to-midpoint part and the midpoint-to-relative part by correctly assuming that the midpoint reference frame is in a state of absolute rest. The solution for each part can be combined to form a complete solution for the midpoint theory of relativity.

1.2.2   The midpoint relative-velocity constraint

The midpoint theory of relativity constrains the relative velocity between the event reference frame and the relative reference frame to be less than **twice the speed of light**, which is referred to as the **midpoint relative-velocity constraint**. This can be seen by the **midpoint relativistic factor**, which is at the mathematical heart of the theory. The midpoint relativistic factor can be written as:

$$\gamma_m = \frac{1}{\sqrt{1 - \frac{v^2}{4c^2}}} \qquad (1\text{-}6)$$

where γ_m is the midpoint relativistic factor (or midpoint gamma), v is the relative velocity between the two standard reference frames, and c is the speed of light in a vacuum. The relative velocity can be written as a fraction of the speed of light:

$$v = \beta_m c \qquad (1\text{-}7)$$

where v is the relative velocity between the event reference frame and the relative reference frame, β_m represents a fraction, and c is the speed of light in a vacuum. The midpoint relativistic factor can be rewritten in terms of the fraction β_m by substituting equation (1-7) into equation (1-6) to obtain:

$$\gamma_m = \frac{1}{\sqrt{1 - \frac{(\beta_m c)^2}{4c^2}}} = \frac{1}{\sqrt{1 - \frac{\beta_m^2}{4}}} \qquad (1\text{-}8)$$

where γ_m is the midpoint relativistic factor, β_m is the fraction, and c is the speed of light in a vacuum.

When the relative velocity between the two standard reference frames is equal to twice the speed of light ($\beta_m = 2$), the midpoint relativistic factor is infinite ($\gamma_m = 1/0 = \infty$). In addition, when the relative velocity between the two standard reference frames is greater than twice the speed of light ($\beta_m > 2$), the term in the square root becomes negative, which results in an imaginary midpoint relativistic factor. The midpoint relative-velocity constraint for the midpoint theory of relativity restricts the relative velocity between the two standard reference frames to be less than twice the speed of light, and can be written in terms of the fraction β_m as:

$$0 \leq \beta_m < 2 \qquad (1\text{-}9)$$

Alternatively, the midpoint relative-velocity constraint can be rewritten in terms of the relative velocity by substituting $\beta_m = v/c$ from equation (1-7) into equation (1-9) to give:

$$0 \leq v < 2c \qquad (1\text{-}10)$$

where v is the relative velocity between the two standard reference frames, and c is the speed of light in a vacuum.

The relative-velocity range for the midpoint theory of relativity is **twice** the relative-velocity range for the special theory of relativity. The relative-velocity range extension allows midpoint relativity to model two standard reference frames travelling in opposite directions, where each standard reference frame travels at more than half the speed of light, but neither of the standard reference frames violate the light-speed constraint.

1.3 Acknowledgement

The special theory of relativity is over one-hundred years old. Descriptions and thought experiments about the theory are well developed. This book has used the excellent textbook of Young and Freedman (2000) as a guide.

2. Time Intervals

The measurement of **time intervals** between two events is dependent on the observer's reference frame. If the events occur in the observer's reference frame, all the normal laws of physics apply. In contrast, if the events occur in another reference frame, the laws of relativity apply.

2.1 Thought Experiment

A thought experiment is used to demonstrate the measurement of time intervals from another reference frame. The thought experiment consists of a light clock, where a light source emits a light pulse that travels to a mirror, where it is reflected and then returns to the light source.

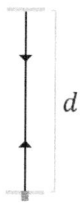

Figure 2-1. Thought experiment to measure time intervals

An observer measures the time interval for the light pulse to travel from the light source to the mirror and then back to the light source. Figure 2-1 displays the thought experiment. Using the normal laws of physics, the time it takes for the light pulse to travel from the light source to the mirror and then back to the light source is:

$$\Delta t = \frac{2d}{c} \tag{2-1}$$

where Δt is the time interval, d is the distance between the light source and the mirror, and c is the speed of light. The thought experiment is used to show how time intervals are measured from the perspective of both the special theory of relativity and the midpoint theory of relativity.

2.2   Principle of Relativity

The principle of relativity states that all the normal laws of physics are the same in all reference frames. Both special relativity and midpoint relativity use observers in their respective thought experiments.

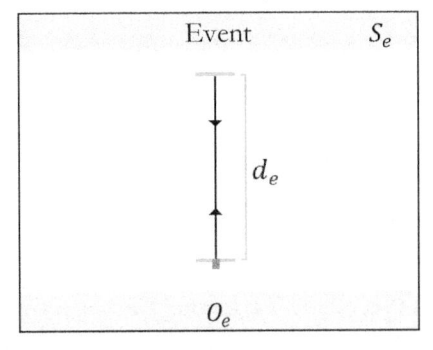

Figure 2-2. Time interval measurement for the event observer

The events for the thought experiment occur in the event reference frame, which contains an event observer. Figure 2-2 displays the thought experiment for the event observer O_e in the event reference frame S_e. All the normal laws of physics apply because the events for the time interval occur in the event observer's reference frame. Thus, equation (2-1) can be used by the event observer to measure the time interval for the light pulse to travel from the light source to the mirror and then back to the light source by:

$$\Delta t_e = \frac{2d_e}{c} \tag{2-2}$$

where Δt_e is the time interval as measured by the event observer, d_e is the distance between the light source and the mirror as measured by the event observer, and c is the speed of light.

2.3 Special Relativity

The special theory of relativity is used to measure **time intervals** of events that occur in another reference frame from the observer. More specifically, the events for the time interval occur in the event reference frame and are measured by the relative observer in the relative reference frame. Special relativity assumes that one of the reference frames is in a state of absolute rest.

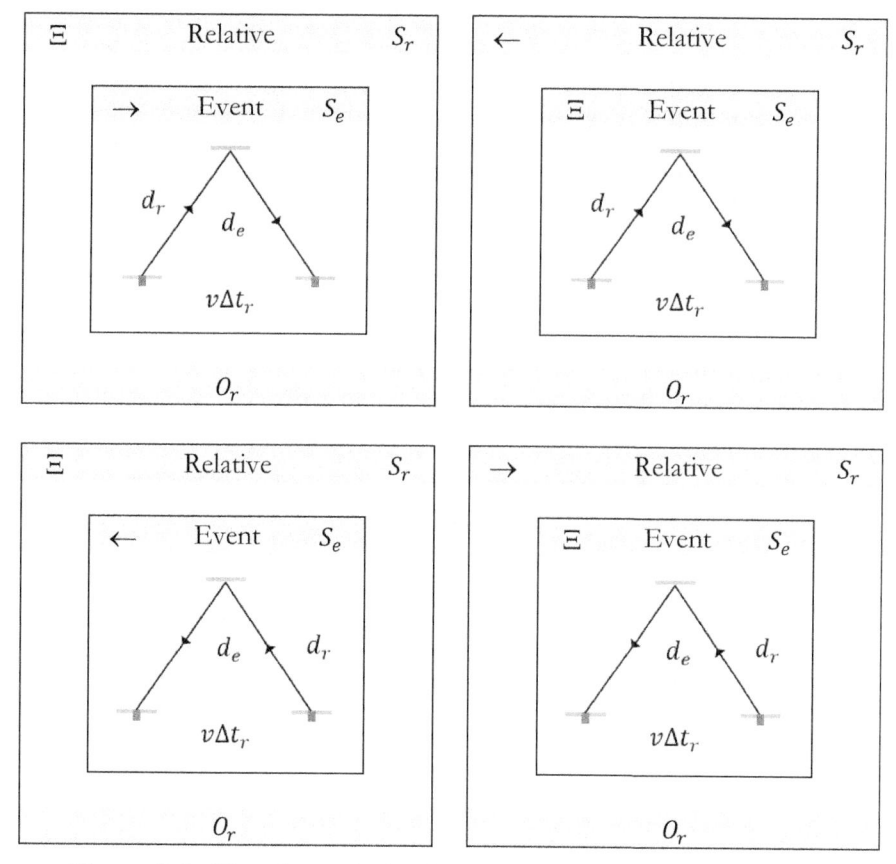

Figure 2-3. Time intervals for the four scenarios of special relativity

The special theory of relativity has four different scenarios for the two reference frames, which are discussed in Section 1.1.1. Figure 2-3 displays the thought experiment for all four scenarios. For each scenario, one reference frame is in a state of absolute rest.

The events for the time interval occur in the event reference frame. By the time the light pulse from the light source hits the mirror, the event reference frame has moved a distance $\frac{1}{2}v\Delta t_r$ with respect to the relative reference frame in all four scenarios. The direction of the uniform motion between the reference frames can be ignored, as the calculations are the same for each of the four scenarios. Pythagoras's Theorem can be used to calculate the distance travelled by the light pulse during the time interval by:

$$d_r^2 = d_e^2 + \left(\tfrac{1}{2}v\Delta t_r\right)^2 \tag{2-3}$$

where d_r is the distance travelled by the light pulse as measured by the relative observer, d_e is the distance travelled by the light pulse as measured by the event observer, Δt_r is the time interval as measured by the relative observer, and v is the relative velocity between the event reference frame and the relative reference frame.

The speed of light in a vacuum is the same in all reference frames. Equation (2-1) can be used to substitute $d_r = \tfrac{1}{2}c\Delta t_r$ and $d_e = \tfrac{1}{2}c\Delta t_e$ into equation (2-3) and collect like terms to produce:

$$\left(1 - \frac{v^2}{c^2}\right)\Delta t_r^2 = \Delta t_e^2 \tag{2-4}$$

Both sides of equation (2-4) can be divided by $1 - v^2/c^2$, and taken the square root of, to obtain:

$$\boxed{\Delta t_r = \frac{\Delta t_e}{\sqrt{1 - \frac{v^2}{c^2}}} = \gamma_s \Delta t_e} \tag{2-5}$$

where Δt_r is the time interval as measured by the relative observer, Δt_e is the time interval as measured by the event observer, v is the relative velocity between the event reference frame and the relative reference frame, c is the speed of light, and γ_s is the **special relativistic factor** (or **special gamma**) given by:

$$\gamma_s = \frac{1}{\sqrt{1 - \frac{v^2}{c^2}}} \tag{2-6}$$

The value for the special relativistic factor in equation (2-5) is always greater than or equal to one, since the relative velocity is always less than the speed of light, due to the special relative-velocity constraint. The **relative** observer measures a longer time interval compared to the **event** observer's measurement of the same time interval. This increased time interval as measured by the relative observer is referred to as **special time dilation**.

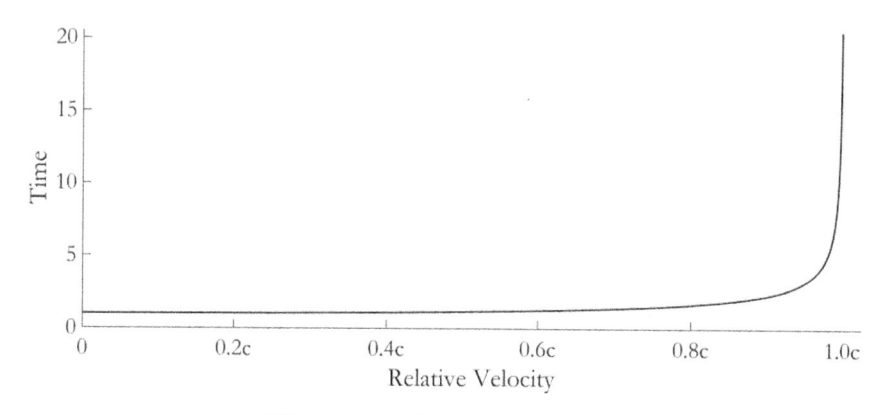

Figure 2-4. Special time dilation

Figure 2-4 displays the time-interval values as measured by the relative observer against the relative-velocity values. To aid interpretation, all time-interval values are measured using equation (2-5) with a one-year time interval in the event reference frame ($\Delta t_e = 1$). The relative-velocity values are fractions of the speed of light. Table 2-1 reports a sample of the time-interval values against the relative-velocity values.

Table 2-1. Special time dilation

Δt_e	Relative Velocity							
	0.00c	0.20c	0.40c	0.60c	0.80c	0.90c	0.95c	0.99c
1.00	1.00	1.02	1.09	1.25	1.67	2.29	3.20	7.09

The closer the relative velocity is to the speed of light, the larger the time dilation measured by the relative observer. For example, when the relative velocity between the event reference frame and the relative reference frame is $0.99c$, a one-year time interval in the event reference frame is measured as 7.09 years by the relative observer in the relative reference frame.

2.4 Midpoint Relativity

The midpoint theory of relativity does **not** assume that either the event reference frame or the relative reference frame is in a state of absolute rest. In a system that consists of two **standard** reference frames in uniform motion, the midpoint is the only point that is in a state of absolute rest. The midpoint is the equidistant point between the event in the event reference frame and the relative observer in the relative reference frame. Midpoint relativity creates a **phantom** midpoint reference frame with a phantom midpoint observer. The midpoint reference frame is always in a state of absolute rest.

The thought experiment used to measure time intervals for the special theory of relativity is reused to measure time intervals for the midpoint theory of relativity. The thought experiment consists of a light clock, where a light source emits a light pulse that travels to a mirror, where it is reflected and then returns to the light source. To understand the measurement of time intervals using the midpoint theory of relativity, the thought experiment can be divided into two parts: an **event-to-midpoint part** and a **midpoint-to-relative part**. The special theory of relativity can be applied separately to both parts. The measurement of time intervals for each part can be combined to form a complete measurement of time intervals for the midpoint theory of relativity.

2.4.1 The event-to-midpoint part

The event-to-midpoint part assumes that the midpoint reference frame is in a state of absolute rest and the event reference frame is travelling with uniform motion either to the left or to the right.

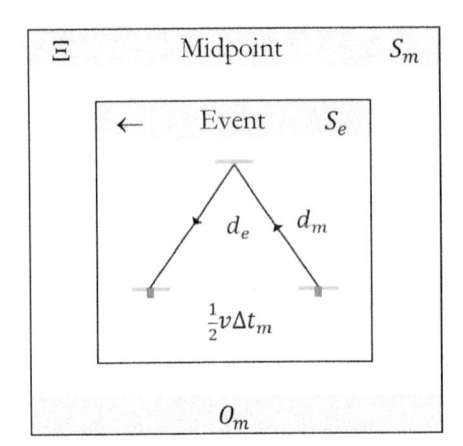

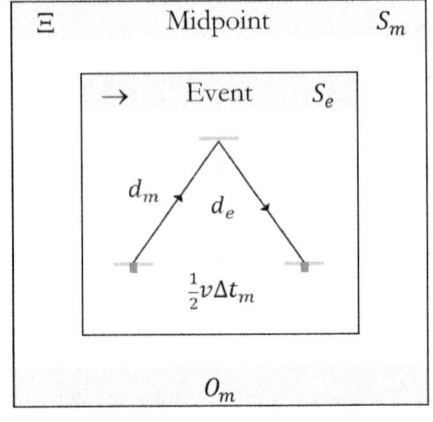

Figure 2-5. Time intervals for the two scenarios of the event-to-midpoint part

The **event-to-midpoint part** of the midpoint theory of relativity has two different scenarios, which are discussed in Section 1.2.1. Figure 2-5 displays the thought experiment for both scenarios. For each scenario, the midpoint reference frame is in a state of absolute rest.

The events for the time interval occur in the event reference frame that is moving with speed $\frac{1}{2}v$ with respect to the midpoint reference frame, where v is the relative velocity between the event reference frame and the relative reference frame. The direction of the uniform motion between the reference frames can be ignored, as the calculations are the same for both scenarios. By the time the light pulse from the light source hits the mirror, the event reference frame has moved a distance $\frac{1}{4}v\Delta t_m$ with respect to the midpoint reference frame. Pythagoras's Theorem can be used to calculate the distance travelled by the light pulse during the time interval by:

$$d_m^2 = d_e^2 + \left(\tfrac{1}{4}v\Delta t_m\right)^2 \tag{2-7}$$

where d_m is the distance travelled by the light pulse as measured by the midpoint observer, d_e is the distance travelled by the light pulse as measured by the event observer, Δt_m is the time interval as measured by the midpoint observer, and v is the relative velocity between the event reference frame and the relative reference frame.

The speed of light in a vacuum is the same in all reference frames. Equation (2-1) can be used to substitute $d_m = \tfrac{1}{2}c\Delta t_m$ and $d_e = \tfrac{1}{2}c\Delta t_e$ into equation (2-7) and collect like terms to produce:

$$\left(1 - \frac{v^2}{4c^2}\right)\Delta t_m^2 = \Delta t_e^2 \tag{2-8}$$

Both sides of equation (2-8) can be divided by $1 - v^2/4c^2$, and taken the square root of, to obtain:

$$\boxed{\Delta t_m = \frac{\Delta t_e}{\sqrt{1 - \dfrac{v^2}{4c^2}}} = \gamma_m \Delta t_e} \tag{2-9}$$

where Δt_m is the time interval as measured by the midpoint observer, Δt_e is the time interval as measured by the event observer, v is the relative velocity between the event reference frame and the relative reference frame, c is the speed of light, and γ_m is the **midpoint relativistic factor** (or **midpoint gamma**) given by:

$$\gamma_m = \frac{1}{\sqrt{1 - \dfrac{v^2}{4c^2}}} \tag{2-10}$$

The value for the midpoint relativistic factor in equation (2-9) is always greater than or equal to one, since the relative velocity is always less than twice the speed of light, due to the midpoint relative-velocity constraint. The **midpoint** observer measures a longer time interval compared to the **event** observer's measurement of the same time interval. This increased time interval as measured by the midpoint observer is referred to as **event-to-midpoint time dilation**.

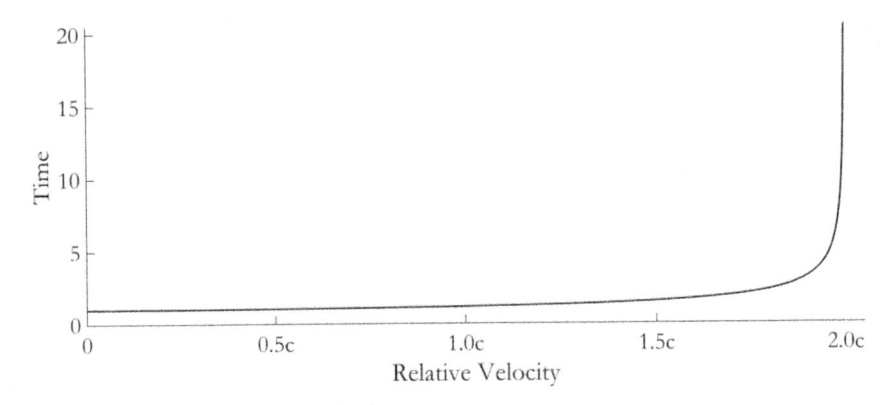

Figure 2-6. Event-to-midpoint time dilation

Figure 2-6 displays the time-interval values as measured by the midpoint observer against the relative-velocity values. To aid interpretation, all time-interval values are measured using equation (2-9) with a one-year time interval in the event reference frame ($\Delta t_e = 1$). The relative-velocity values are fractions of the speed of light. Table 2-2 reports a sample of the time-interval values against the relative-velocity values.

Table 2-2. Event-to-midpoint time dilation

Δt_e	Relative Velocity							
	0.00c	0.40c	0.80c	1.20c	1.60c	1.80c	1.90c	1.98c
1.00	1.00	1.02	1.09	1.25	1.67	2.29	3.20	7.09

The closer the relative velocity is to twice the speed of light, the larger the time dilation measured by the midpoint observer. For example, when the relative velocity between the event reference frame and the relative reference frame is 1.98c, a one-year time interval in the event reference frame is measured as 7.09 years by the midpoint observer in the midpoint reference frame.

2.4.2 The midpoint-to-relative part

The midpoint-to-relative part assumes that the midpoint reference frame is in a state of absolute rest and the relative reference frame is travelling with uniform motion either to the left or to the right.

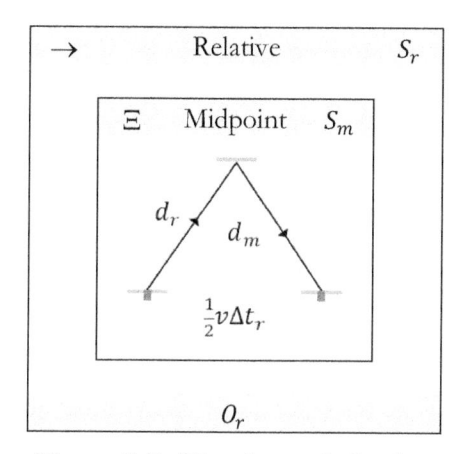

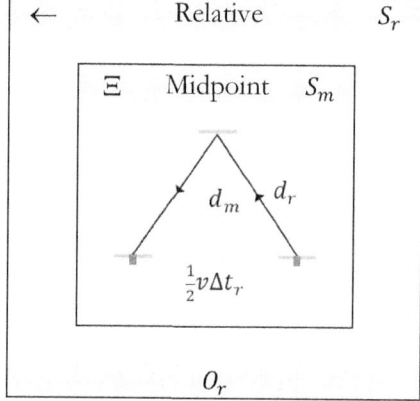

Figure 2-7. Time intervals for the two scenarios of the midpoint-to-relative part

The **midpoint-to-relative part** of the midpoint theory of relativity has two different scenarios, which are discussed in Section 1.2.1. Figure 2-7 displays the thought experiment for both scenarios. For each scenario, the midpoint reference frame is in a state of absolute rest.

The events for the time interval occur in the midpoint reference frame, with the relative reference frame moving with speed $\frac{1}{2}v$ with respect to the midpoint reference frame, where v is the relative velocity between the event reference frame and the relative reference frame. The direction of the uniform motion between the reference frames can be ignored, as the calculations are the same for both scenarios. By the time the light pulse from the light source hits the mirror, the relative reference frame has moved a distance $\frac{1}{4}v\Delta t_r$ with respect to the midpoint reference frame. Pythagoras's Theorem can be used to calculate the distance travelled by the light pulse during the time interval by:

$$d_r^2 = d_m^2 + \left(\frac{1}{4}v\Delta t_r\right)^2 \qquad (2\text{-}11)$$

where d_r is the distance travelled by the light pulse as measured by the relative observer, d_m is the distance travelled by the light pulse as measured by the midpoint observer, Δt_r is the time interval as measured by the relative observer, and v is the relative velocity between the event reference frame and the relative reference frame.

The speed of light in a vacuum is the same in all reference frames. Equation (2-1) can be used to substitute $d_r = \frac{1}{2}c\Delta t_r$ and $d_m = \frac{1}{2}c\Delta t_m$ into equation (2-11) and collect like terms to produce:

$$\left(1 - \frac{v^2}{4c^2}\right)\Delta t_r^2 = \Delta t_m^2 \qquad (2\text{-}12)$$

Both sides of equation (2-12) can be divided by $1 - v^2/4c^2$, and taken the square root of, to obtain:

$$\boxed{\Delta t_r = \frac{\Delta t_m}{\sqrt{1 - \dfrac{v^2}{4c^2}}} = \gamma_m \Delta t_m} \qquad (2\text{-}13)$$

where Δt_r is the time interval as measured by the relative observer, Δt_m is the time interval as measured by the midpoint observer, v is the relative velocity between the event reference frame and the relative reference frame, c is the speed of light, and γ_m is the midpoint relativistic factor.

The value for the midpoint relativistic factor in equation (2-13) is always greater than or equal to one, since the relative velocity is always less than twice the speed of light, due to the midpoint relative-velocity constraint. The **relative observer measures a longer time interval compared to the midpoint observer's measurement of the same time interval.** This increased time interval as measured by the relative observer is referred to as **midpoint-to-relative time dilation**.

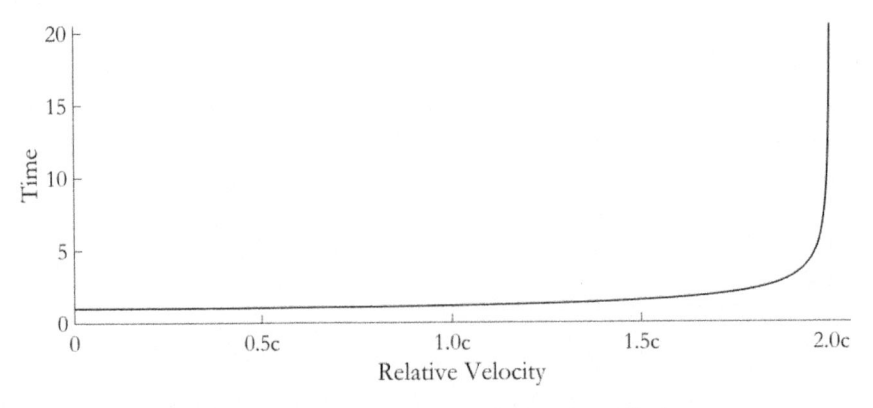

Figure 2-8. Midpoint-to-relative time dilation

Figure 2-8 displays the time-interval values as measured by the relative observer against the relative-velocity values. To aid interpretation, all time-interval values are measured using equation (2-13) with a one-year time interval in the midpoint reference frame ($\Delta t_m = 1$). The relative-velocity values are fractions of the speed of light. Table 2-3 reports a sample of the time-interval values against the relative-velocity values.

Table 2-3. Midpoint-to-relative time dilation

Δt_m	Relative Velocity							
	0.00c	0.40c	0.80c	1.20c	1.60c	1.80c	1.90c	1.98c
1.00	1.00	1.02	1.09	1.25	1.67	2.29	3.20	7.09

The closer the relative velocity is to twice the speed of light, the larger the time dilation measured by the relative observer. For example, when the relative velocity between the event reference frame and the relative reference frame is 1.98c, a one-year time interval in the midpoint reference frame is measured as 7.09 years by the relative observer in the relative reference frame.

2.4.3 The complete solution

The **midpoint theory of relativity** divides the thought experiment into two parts: an event-to-midpoint part and a midpoint-to-relative part. The previous sections applied the special theory of relativity to each part. In this section, the measurements for time interval for both parts are combined to derive a measurement of time intervals for the midpoint theory of relativity. Table 2-4 reports the formula used to measure time intervals for both parts of midpoint relativity.

Table 2-4. Formula comparison for measuring time intervals for both parts of midpoint relativity

Event-to-Midpoint	Midpoint-to-Relative
$\Delta t_m = \gamma_m \Delta t_e$	$\Delta t_r = \gamma_m \Delta t_m$

The event-to-midpoint part assumes that the midpoint reference frame is in a state of absolute rest and the event reference frame is in uniform motion. The time interval formula for the midpoint observer is $\Delta t_m = \gamma_m \Delta t_e$ from equation (2-9). The midpoint-to-relative part assumes that the midpoint reference frame is in a state of absolute rest and the relative reference frame is in uniform motion. The time interval formula for the relative observer is $\Delta t_r = \gamma_m \Delta t_m$ from equation (2-13).

The time interval measurements for both parts can be combined by substituting equation (2-9) into equation (2-13) to obtain:

$$\Delta t_r = \gamma_m^2 \Delta t_e \qquad (2\text{-}14)$$

where Δt_r is the time interval as measured by the relative observer, γ_m is the midpoint relativistic factor, and Δt_e is the time interval as measured by the event observer.

The **squared** value for the midpoint relativistic factor in equation (2-14) is always greater than or equal to one, since the relative velocity is always less than twice the speed of light, due to the midpoint relative-velocity constraint. The **relative** observer measures a longer time interval compared to the **event** observer's measurement of the same time interval. This increased time interval as measured by the relative observer is referred to as **midpoint time dilation**.

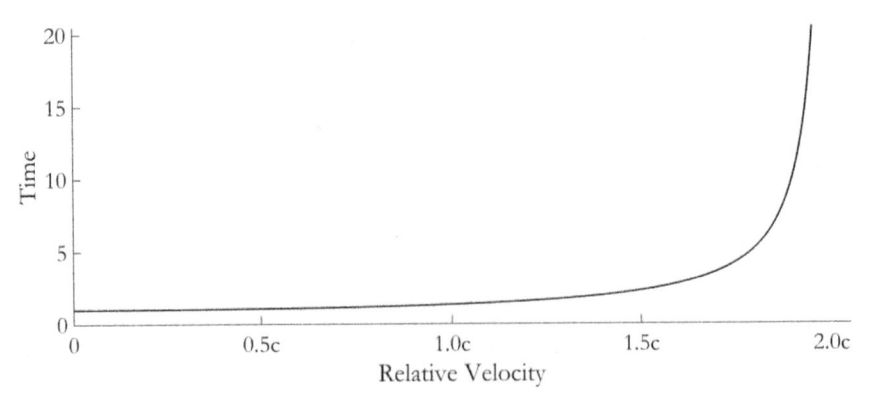

Figure 2-9. Midpoint time dilation

Figure 2-9 displays the time-interval values as measured by the relative observer against the relative-velocity values. To aid interpretation, all time-interval values are measured using equation (2-14) with a one-year time interval in the event reference frame ($\Delta t_e = 1$). The relative-velocity values are fractions of the speed of light. Table 2-5 reports a sample of the time-interval values against the relative-velocity values.

Table 2-5. Midpoint time dilation

Δt_e	Relative Velocity							
	0.00c	0.40c	0.80c	1.20c	1.60c	1.80c	1.90c	1.98c
1.00	1.00	1.04	1.19	1.56	2.78	5.26	10.26	50.25

The closer the relative velocity is to the speed of light, the larger the time dilation measured by the relative observer. For example, when the relative velocity between the event reference frame and the relative reference frame is 1.98c, a one-year time interval in the event reference frame is measured as 50.25 years by the relative observer in the relative reference frame.

2.5 Relativity Comparison

This section compares the formula for the measurement of time intervals for both the special theory of relativity and the midpoint theory of relativity. Table 2-6 reports the formula for the relative observer measuring time intervals between events that occur in the event reference frame.

Table 2-6. Formula comparison for measuring time intervals

Special Relativity	Midpoint Relativity
$\Delta t_r = \gamma_s \Delta t_e$	$\Delta t_r = \gamma_m^2 \Delta t_e$

The relativistic factors drive the magnitude of the time dilations. The time interval in the event reference frame is multiplied by the squared midpoint relativistic factor in the midpoint theory of relativity, compared to the single special relativistic factor in the special theory of relativity. This difference results in significantly different outcomes for both theories of relativity, especially when the relative velocity is greater than half the speed of light.

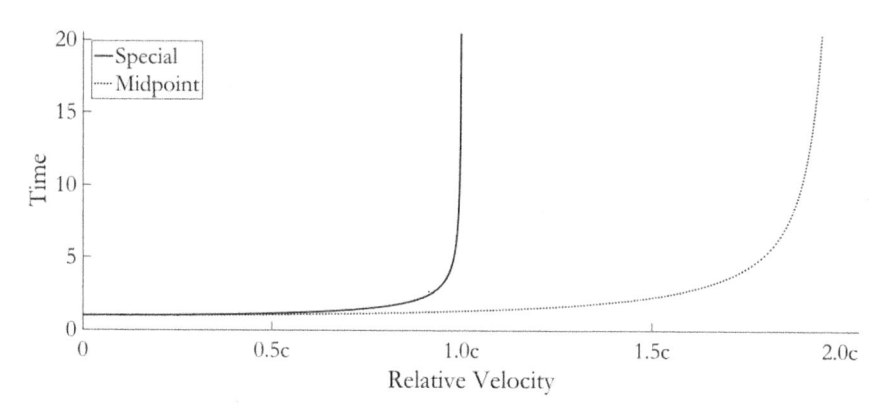

Figure 2-10. Time dilation comparison

Figure 2-10 displays the time-interval values as measured by the relative observer against the relative-velocity values, for both special relativity and midpoint relativity. To aid interpretation, the time-interval values are measured using equation (2-5) for special relativity and equation (2-14) for midpoint relativity, with a one-year time interval in the event reference frame ($\Delta t_e = 1$). The relative-velocity values are fractions of the speed of light. Table 2-7 reports a sample of the time-interval values against the relative-velocity values for both special relativity (SR) and midpoint relativity (MR).

Table 2-7. Time dilation comparison

	Δt_e	\multicolumn{7}{c}{Relative Velocity}
		0.00c	0.50c	0.90c	0.95c	0.99c	1.80c	1.90c	1.98c
SR	1.00	1.00	1.15	2.29	3.20	7.09			
MR	1.00	1.00	1.07	1.25	1.29	1.32	5.26	10.26	50.25

The special theory of relativity is unable to measure time intervals in another reference frame when the relative velocity between the two standard reference frames is greater than the speed of light, due to the special relative-velocity constraint. In contrast, the midpoint theory of relativity has a relative-velocity constraint of less than **twice** the speed of light, due to the midpoint relative-velocity constraint.

When the relative velocity is less than the speed of light, the special theory of relativity overestimates time dilation compared to the midpoint theory of relativity, with most of the overestimation occurring when the relative velocity is close to the speed of light. For example, when the relative velocity between the event reference frame and the relative reference frame is $0.99c$, a one-year time interval in the event reference frame is measured by the relative observer as 7.09 years using the special theory of relativity, compared to 1.32 years using the midpoint theory of relativity.

3. Length of Objects

The measurement of the **length of objects** (or distances) is dependent on the observer's reference frame. If an object is located in the observer's reference frame, all the normal laws of physics apply. In contrast, if an object is located in another reference frame, the laws of relativity apply.

3.1 Thought Experiment

A thought experiment is used to demonstrate the measurement of the length of objects from another reference frame. The thought experiment consists of a ruler with a light source at one end and a mirror at the other end. The light source emits a light pulse that travels to a mirror, where it is reflected and then returns to the light source.

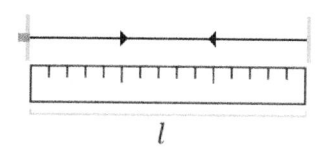

Figure 3-1. Thought experiment to measure the length of objects

An observer measures the time interval for the light pulse to travel from the light source to the mirror and then back to the light source. Figure 3-1 displays the thought experiment. Using the normal laws of physics, the total time it takes for the light pulse to travel two lengths of the ruler (back and forth) is:

$$\Delta t = \frac{2l}{c} \tag{3-1}$$

where Δt is the time interval for the light pulse to travel from the light source to the mirror and back, l is the length of the ruler, and c is the speed of light. This thought experiment is used to show how the length of objects are measured from the perspective of both the special theory of relativity and the midpoint theory of relativity.

3.2 Principle of Relativity

The principle of relativity states that all the normal laws of physics are the same in all reference frames. Both special relativity and midpoint relativity use observers in their respective thought experiments.

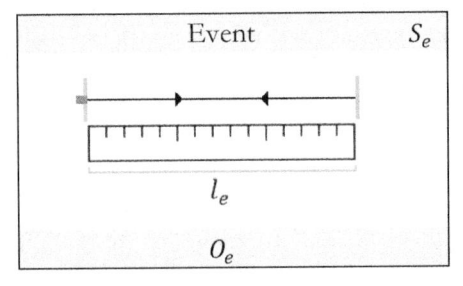

Figure 3-2. Length measurement for the event observer

The ruler for the thought experiment is located in the event reference frame, which contains the event observer. Figure 3-2 displays the thought experiment for the event observer O_e in the event reference frame S_e. All the normal laws of physics apply because the object is located in the event observer's reference frame. Thus, equation (3-1) can be used to measure the total time taken for the light pulse to travel two lengths of the ruler (back and forth):

$$\Delta t_e = \frac{2l_e}{c} \qquad (3\text{-}2)$$

where Δt_e is the time interval for the light pulse to travel from the light source to the mirror and then is reflected back to the source as measured by the event observer, l_e is the length of the ruler as measured by the event observer, and c is the speed of light.

In the thought experiment, the light pulse travels in **two directions**: the **source-to-mirror direction** and the **mirror-to-source direction**. The two directions are important for measuring the length of an object in another reference frame. The source-to-mirror direction is when the light pulse travels from the light source to the mirror, whilst moving **away from** the observer in another reference frame. The mirror-to-source direction is when the light pulse travels from the mirror back to the light source, whilst moving **towards** the observer in another reference frame. It should be noted that the overall result would be the same if the light source and mirror were at alternate ends.

3.3 Special Relativity

The special theory of relativity measures the **length of objects** that are located in another reference frame from the observer. More specifically, the object is located in the event reference frame and is measured by the relative observer in the relative reference frame. Special relativity assumes that one of the reference frames is in a state of absolute rest.

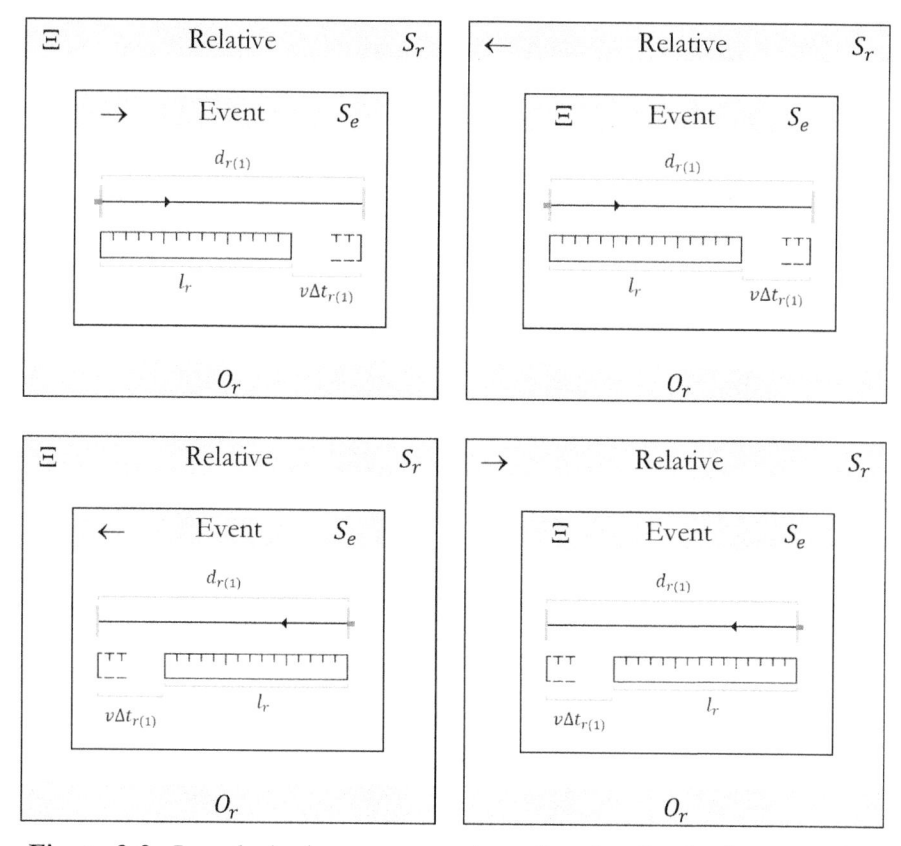

Figure 3-3. Lengths in the source-to-mirror direction for the four scenarios of special relativity

For the **source-to-mirror direction**, the special theory of relativity has four different scenarios for the two reference frames, which are discussed in Section 1.1.1. Figure 3-3 displays the thought experiment for all four scenarios. For each scenario, one reference frame is assumed to be in a state of absolute rest.

The ruler within the event reference frame moves **away from** the relative reference frame. The light pulse also moves with speed c **away from** the relative reference frame. In the time taken for the light pulse to travel from the light source to the mirror, the relative reference frame has moved a distance $v\Delta t_{r(1)}$ with respect to the event reference frame. The distance travelled by the light pulse in the time interval $\Delta t_{r(1)}$ is given by:

$$d_{r(1)} = l_r + v\Delta t_{r(1)} \tag{3-3}$$

where $d_{r(1)}$ is the distance travelled by the light pulse in the time interval $\Delta t_{r(1)}$ as measured by the relative observer, l_r is the length of the ruler as measured by the relative observer, and v is the relative velocity between the event reference frame and the relative reference frame.

The light pulse travels with speed c, so equation (2-1) can be used to measure the distance $d_{r(1)}$ travelled by the light pulse in the time interval $\Delta t_{r(1)}$ to give:

$$d_{r(1)} = c\Delta t_{r(1)} \tag{3-4}$$

Equation (3-3) and equation (3-4) can be combined to eliminate $d_{r(1)}$ and solved in terms of $\Delta t_{r(1)}$ to give:

$$\Delta t_{r(1)} = \frac{l_r}{c - v} \tag{3-5}$$

where $\Delta t_{r(1)}$ is the time taken for the light pulse to travel from the light source to the mirror as measured by the relative observer, l_r is the length of the ruler as measured by the relative observer, v is the relative velocity between the event reference frame and the relative reference frame, and c is the speed of light.

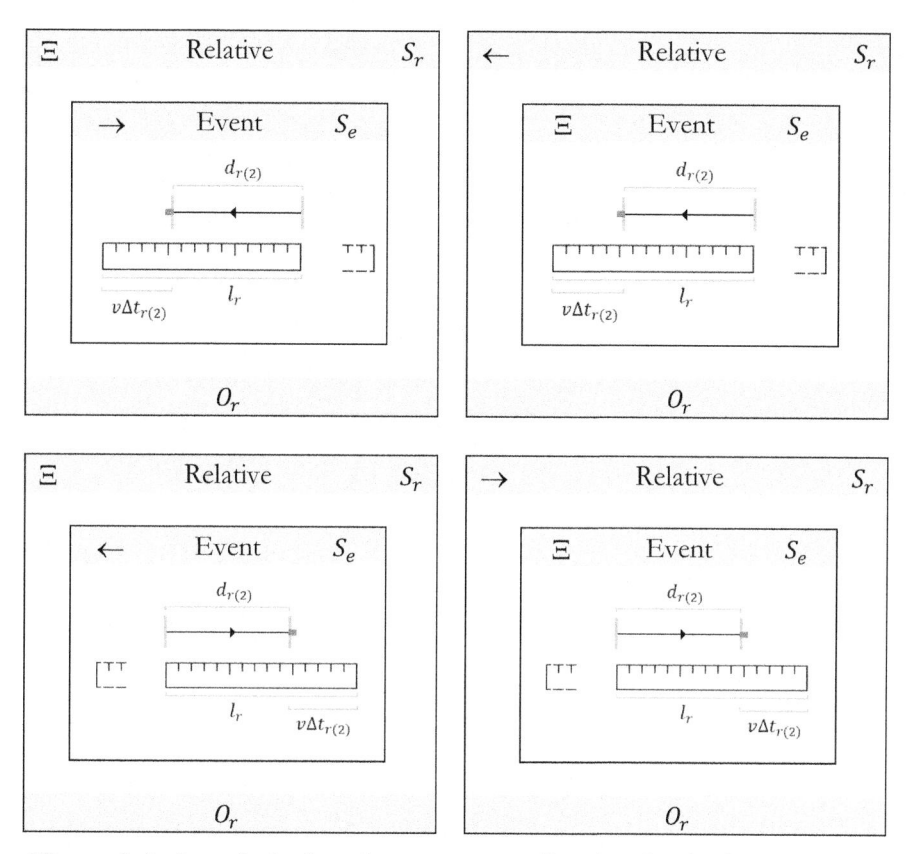

Figure 3-4. Lengths in the mirror-to-source direction for the four scenarios of special relativity

For the **mirror-to-source direction**, the special theory of relativity has four different scenarios for the two reference frames, which are discussed in Section 1.1.1. Figure 3-4 displays the thought experiment for all four scenarios. For each scenario, one reference frame is assumed to be in a state of absolute rest.

The ruler within the event reference frame still moves **away from** the relative reference frame. However, the light pulse travels with speed c **towards** the relative reference frame. In the time taken for the light pulse to travel from the mirror back to the light source, the relative reference frame has moved a distance $v\Delta t_{r(2)}$ with respect to the event reference frame. The distance travelled by the light pulse in the time interval $\Delta t_{r(2)}$ is given by:

$$d_{r(2)} = l_r - v\Delta t_{r(2)} \tag{3-6}$$

where $d_{r(2)}$ is the distance travelled by the light pulse in the time interval $\Delta t_{r(2)}$ as measured by the relative observer, l_r is the length of the ruler as measured by the relative observer, and v is the relative velocity between the event reference frame and the relative reference frame.

The light pulse travels with speed c, so equation (2-1) can be used to measure the distance d_r travelled by the light pulse in the time interval $\Delta t_{r(2)}$ to give:

$$d_{r(2)} = c\Delta t_{r(2)} \tag{3-7}$$

Equation (3-6) and equation (3-7) can be combined to eliminate $d_{r(2)}$ and solved in terms of $\Delta t_{r(2)}$ to give:

$$\Delta t_{r(2)} = \frac{l_r}{c + v} \tag{3-8}$$

where $\Delta t_{r(2)}$ is the time taken for the light pulse to travel from the mirror back to the light source as measured by the relative observer, l_r is the length of the ruler as measured by the relative observer, v is the relative velocity between the event reference frame and the relative reference frame, and c is the speed of light.

The **total time** taken for the light pulse to travel from the light source to the mirror and then back to the light source is the sum of the time taken in the source-to-mirror direction $\Delta t_{r(1)}$ and the time taken in the mirror-to-source direction $\Delta t_{r(2)}$:

$$\Delta t_r = \frac{(c+v)l_r + (c-v)l_r}{c^2\left(1 - \frac{v^2}{c^2}\right)} = \frac{2\gamma_s^2 l_r}{c} \tag{3-9}$$

where Δt_r is the time interval for the light pulse to travel from the light source to the mirror and then back to the light source as measured by the relative observer, and γ_s is the special relativistic factor.

Equation (3-2) for Δt_e can be substituted into the special time dilation equation $\Delta t_r = \gamma_s \Delta t_e$ in (2-5) to give:

$$\Delta t_r = \frac{2\gamma_s l_e}{c} \tag{3-10}$$

Finally, equation (3-9) and equation (3-10) can be combined to eliminate Δt_r and solved for l_r to obtain:

$$l_r = \frac{1}{\gamma_s} l_e \qquad (3\text{-}11)$$

where l_r is the length of the ruler as measured by the relative observer, γ_s is the special relativistic factor, and l_e is the length of the ruler as measured by the event observer.

The **inverse** value of the special relativistic factor in equation (3-11) is always less than or equal to one, since the relative velocity is always less than the speed of light, due to the special relative-velocity constraint. The **relative observer** measures a shorter length of the ruler compared to the **event observer**'s length measurement of the same ruler. This shortened length measured by the relative observer is referred to as **special length contraction**.

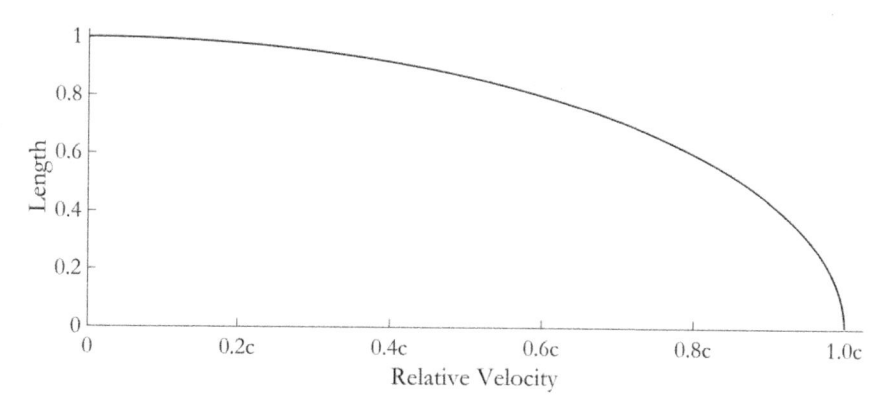

Figure 3-5. Special length contraction

Figure 3-5 displays the length-of-the-ruler values as measured by the relative observer against the relative-velocity values. To aid interpretation, all the length-of-the-ruler values are measured using equation (3-11) with a one-metre ruler in the event reference frame ($l_e = 1$). The relative-velocity values are fractions of the speed of light. Table 3-1 reports a sample of the length-of-the-ruler values against the relative-velocity values.

Table 3-1. Special length contraction

l_e	Relative Velocity							
	0.00c	0.20c	0.40c	0.60c	0.80c	0.90c	0.95c	0.99c
1.00	1.00	0.98	0.92	0.80	0.60	0.44	0.31	0.14

The closer the relative velocity is to the speed of light, the larger the length contraction measured by the relative observer. For example, when the relative velocity between the event reference frame and the relative reference frame is $0.99c$, a one-metre rule in the event reference frame is measured as 0.14 metres by the relative observer in the relative reference frame.

3.4 Midpoint Relativity

The midpoint theory of relativity does **not** assume that either the event reference frame or the relative reference frame is in a state of absolute rest. In a system that consists of two **standard** reference frames in uniform motion, the midpoint is the only point that is in a state of absolute rest. The midpoint is the equidistant point between the object in the event reference frame and the relative observer in the relative reference frame. Midpoint relativity creates a **phantom** midpoint reference frame with a phantom midpoint observer. The midpoint reference frame is always in a state of absolute of rest.

The thought experiment used to measure the length of objects for the special theory of relativity is reused to measure the **length of objects** for the midpoint theory of relativity. The thought experiment consists of a ruler with a light source at one end and a mirror at the other end. The light source emits a light pulse that travels to the mirror, where it is reflected and then returns to the light source. To understand the measurement of the length of objects using the midpoint theory of relativity, the thought experiment can be divided into two parts: an **event-to-midpoint part** and a **midpoint-to-relative part**. The special theory of relativity can then be applied separately to both parts. The measurement of the length of objects for each part can be combined to form a complete measurement of the length of objects for the midpoint theory of relativity.

3.4.1 The event-to-midpoint part

The event-to-midpoint part assumes that the midpoint reference frame is in a state of absolute rest and the event reference frame is travelling with uniform motion either to the left or to the right.

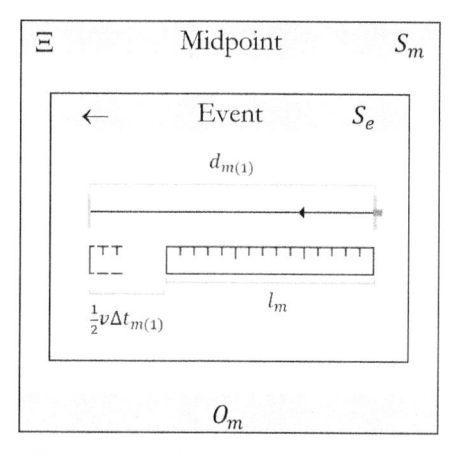

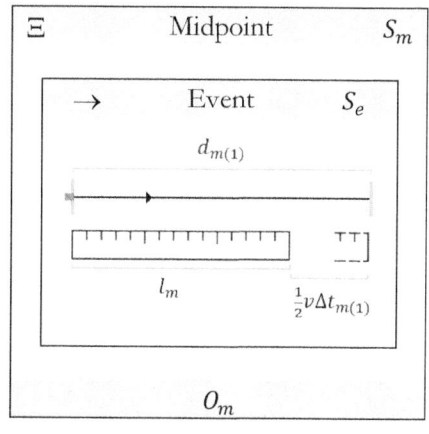

Figure 3-6. Lengths in the source-to-mirror direction for the two scenarios of the event-to-midpoint part

For the **source-to-mirror direction**, the event-to-midpoint part of the midpoint theory of relativity has two different scenarios, which are discussed in Section 1.2.1. Figure 3-6 displays the source-to-mirror direction of the thought experiment for both scenarios. For each scenario, the midpoint reference frame is in a state of absolute rest.

The object is located in the event reference frame that moves with speed $\frac{1}{2}v$ with respect to the midpoint reference frame, where v is the relative velocity between the event reference frame and the relative reference frame. The ruler in the event reference frame moves **away from** the midpoint reference frame. The light pulse also moves with speed c **away from** the midpoint reference frame. In the time taken for the light pulse to travel from the light source to the mirror, the event reference frame has moved a distance $\frac{1}{2}v\Delta t_{m(1)}$ with respect to the midpoint reference frame. The distance travelled by the light pulse in the time interval $\Delta t_{m(1)}$ is given by:

$$d_{m(1)} = l_m + \frac{1}{2}v\Delta t_{m(1)} \tag{3-12}$$

where $d_{m(1)}$ is the distance travelled by the light pulse in the time interval $\Delta t_{m(1)}$ as measured by the midpoint observer, l_m is the length of the ruler as measured by the midpoint observer, and v is the relative velocity between the event reference frame and the relative reference frame.

The light pulse travels with speed c, so equation (2-1) can be used to measure the distance $d_{m(1)}$ travelled by the light pulse in the time interval $\Delta t_{m(1)}$ to give:

$$d_{m(1)} = c\Delta t_{m(1)} \tag{3-13}$$

Equation (3-12) and equation (3-13) can be combined to eliminate $d_{m(1)}$ and solved in terms of $\Delta t_{m(1)}$ to give:

$$\Delta t_{m(1)} = \frac{l_m}{c - \frac{1}{2}v} \qquad (3\text{-}14)$$

where $\Delta t_{m(1)}$ is the time taken for the light pulse to travel from the light source to the mirror as measured by the midpoint observer, l_m is the length of the ruler as measured by the midpoint observer, v is the relative velocity between the event reference frame and the relative reference frame, and c is the speed of light.

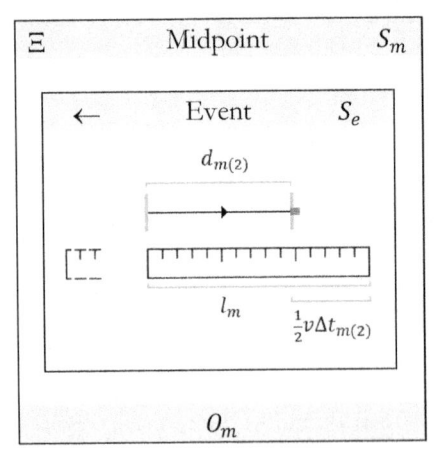

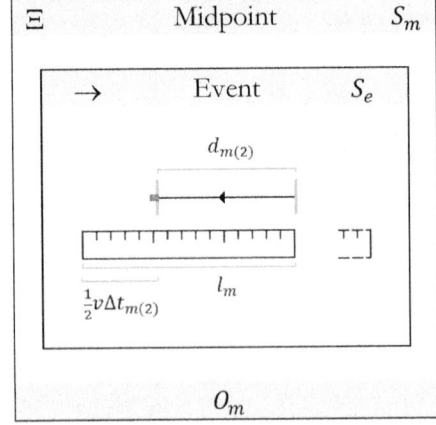

Figure 3-7. Lengths in the mirror-to-source direction for the two scenarios of the event-to-midpoint part

For the **mirror-to-source direction**, the event-to-midpoint part of the midpoint theory of relativity has two different scenarios, which are discussed in Section 1.2.1. Figure 3-7 displays the mirror-to-source direction of the thought experiment for both scenarios. For each scenario, the midpoint reference frame is in a state of absolute rest.

The object is located in the event reference frame that moves with speed $\frac{1}{2}v$ with respect to the midpoint reference frame, where v is the relative velocity between the event reference frame and the relative reference frame. The ruler within the event reference frame still moves **away from** the midpoint reference frame. However, the light pulse travels with speed c **towards** the midpoint reference frame. In the time taken for the light pulse to travel from the mirror back to the light source, the event reference frame has moved a distance $\frac{1}{2}v\Delta t_{m(2)}$ with respect to the midpoint reference frame. The distance travelled by the light pulse in the time interval $\Delta t_{m(2)}$ is given by:

$$d_{m(2)} = l_m - \tfrac{1}{2}v\Delta t_{m(2)} \qquad (3\text{-}15)$$

where $d_{m(2)}$ is the distance travelled by the light pulse in the time interval $\Delta t_{m(2)}$ as measured by the midpoint observer, l_m is the length of the ruler as measured by the midpoint observer, and v is the relative velocity between the event reference frame and the relative reference frame.

The light pulse travels with speed c, so equation (2-1) can be used to measure the distance $d_{m(2)}$ travelled by the light pulse in the time interval $\Delta t_{m(2)}$ to give:

$$d_{m(2)} = c\Delta t_{m(2)} \qquad (3\text{-}16)$$

Equation (3-15) and equation (3-16) can be combined to eliminate $d_{m(2)}$ and solved in terms of $\Delta t_{m(2)}$ to obtain:

$$\Delta t_{m(2)} = \frac{l_m}{c + \tfrac{1}{2}v} \qquad (3\text{-}17)$$

where $\Delta t_{m(2)}$ is the time taken for the light pulse to travel from the mirror back to the light source as measured by the midpoint observer, l_m is the length of the ruler as measured by the midpoint observer, v is the relative velocity between the event reference frame and the relative reference frame, and c is the speed of light.

The **total time** taken for the light pulse to travel from the light source to the mirror and then back to the light source is the sum of the time taken in the source-to-mirror direction $\Delta t_{m(1)}$ and the time taken in the mirror-to-source direction $\Delta t_{m(2)}$:

$$\Delta t_m = \frac{\left(c + \tfrac{1}{2}v\right)l_m + \left(c - \tfrac{1}{2}v\right)l_m}{c^2\left(1 - \dfrac{v^2}{4c^2}\right)} = \frac{2\gamma_m^2 l_m}{c} \qquad (3\text{-}18)$$

where Δt_m is the time interval for the light pulse to travel from the light source to the mirror and then back to the light source as measured by the midpoint observer, and γ_m is the midpoint relativistic factor.

Equation (3-2) for Δt_e can be substituted into the event-to-midpoint time dilation equation $\Delta t_m = \gamma_m \Delta t_e$ in (2-9) to give:

$$\Delta t_m = \frac{2\gamma_m l_e}{c} \qquad (3\text{-}19)$$

Finally, equation (3-18) and equation (3-19) can be combined to eliminate Δt_m and solved for l_m to obtain:

$$l_m = \frac{1}{\gamma_m} l_e \qquad (3\text{-}20)$$

where l_m is the length of the ruler as measured by the midpoint observer, γ_m is the midpoint relativistic factor, and l_e is the length of the ruler as measured by the event observer.

The **inverse** value for the midpoint relativistic factor in equation (3-20) is always less than or equal to one, since the relative velocity is always less than twice the speed of light, due to the midpoint relative-velocity constraint. The **midpoint** observer measures a shorter length of the ruler compared to the **event** observer's length measurement of the same ruler. This shortened length measured by the midpoint observer is referred to as **event-to-midpoint length contraction**.

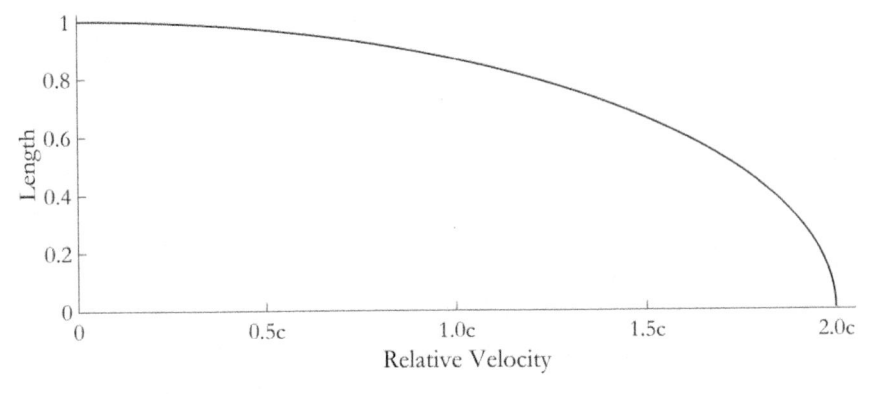

Figure 3-8. Event-to-midpoint length contraction

Figure 3-8 displays the length-of-the-ruler values as measured by the midpoint observer against the relative-velocity values. To aid interpretation, all the length-of-the-ruler values are measured using equation (3-20) with a one-metre ruler in the event reference frame ($l_e = 1$). The relative-velocity values are fractions of the speed of light. Table 3-2 reports a sample of the length-of-the-ruler values against the relative-velocity values.

Table 3-2. Event-to-midpoint length contraction

l_m	Relative Velocity							
	0.00c	0.40c	0.80c	1.20c	1.60c	1.80c	1.90c	1.98c
1.00	1.00	0.98	0.92	0.80	0.60	0.44	0.31	0.14

The closer the relative velocity is to twice the speed of light, the larger the length contraction measured by the midpoint observer. For example, when the relative velocity between the event reference frame and the relative reference frame is 1.98c, a one-metre ruler within the event reference frame is measured as 0.14 metres by the midpoint observer in the midpoint reference frame.

3.4.2 The midpoint-to-relative part

The midpoint-to-relative part assumes that the midpoint reference frame is in a state of absolute rest and the relative reference frame is travelling with uniform motion either to the left or to the right.

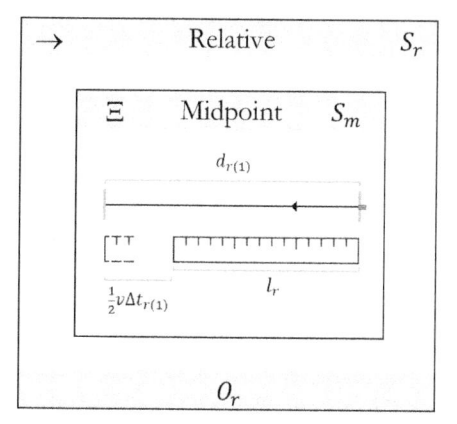

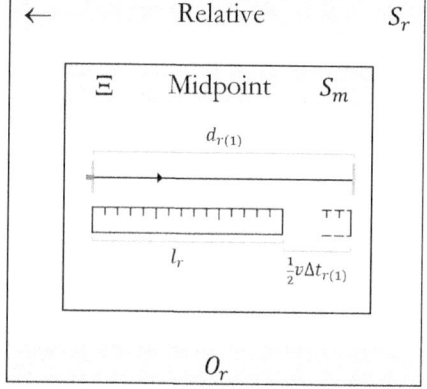

Figure 3-9. Lengths in the source-to-mirror direction for the two scenarios of the midpoint-to-relative part

For the **source-to-mirror direction**, the midpoint-to-relative part of the midpoint theory of relativity has two different scenarios, which are discussed in Section 1.2.1. Figure 3-9 displays the source-to-mirror direction of the thought experiment for both scenarios. For each scenario, the midpoint reference frame is in a state of absolute rest.

The object is located in the midpoint reference frame, with the relative reference frame moving with speed $\frac{1}{2}v$ with respect to the midpoint reference frame, where v is the relative velocity between the event reference frame and the relative reference frame. The ruler within the midpoint reference frame moves **away from** the relative reference frame. The light pulse also moves with speed c **away from** the relative reference frame. In the time taken for the light pulse to travel from the light source to the mirror, the relative reference frame has moved a distance $\frac{1}{2}v\Delta t_{r(1)}$ with respect to the midpoint reference frame. The distance travelled by the light pulse in the time interval $\Delta t_{r(1)}$ is given by:

$$d_{r(1)} = l_r + \frac{1}{2}v\Delta t_{r(1)} \qquad (3\text{-}21)$$

where $d_{r(1)}$ is the distance travelled by the light pulse in the time interval $\Delta t_{r(1)}$ as measured by the relative observer, l_r is the length of the ruler as measured by the relative observer, and v is the relative velocity between the event reference frame and the relative reference frame.

The light pulse travels with speed c, so equation (2-1) can be used to measure the distance $d_{r(1)}$ travelled by the light pulse in the time interval $\Delta t_{r(1)}$ to give:

$$d_{r(1)} = c\Delta t_{r(1)} \qquad (3\text{-}22)$$

Equation (3-21) and equation (3-22) can be combined to eliminate $d_{r(1)}$ and solved in terms of $\Delta t_{r(1)}$ to give:

$$\Delta t_{r(1)} = \frac{l_r}{c - \frac{1}{2}v} \qquad (3\text{-}23)$$

where $\Delta t_{r(1)}$ is the time taken for the light pulse to travel from the light source to the mirror as measured by the relative observer, l_r is the length of the ruler as measured by the relative observer, v is the relative velocity between the event reference frame and the relative reference frame, and c is the speed of light.

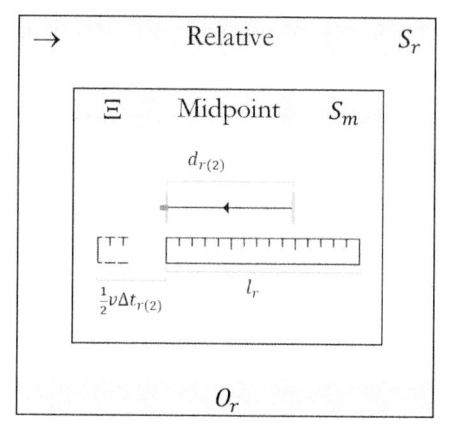

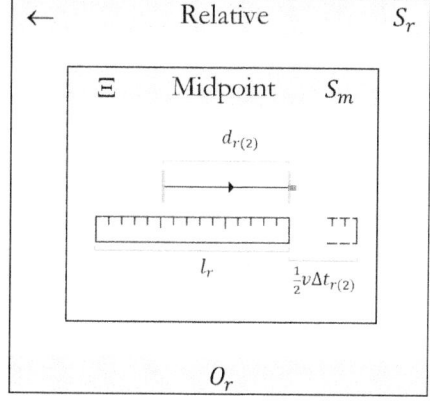

Figure 3-10. Lengths in the mirror-to-source direction for the two scenarios of the midpoint-to-relative part

For the **mirror-to-source direction**, the midpoint-to-relative part of the midpoint theory of relativity has two different scenarios, which are discussed in Section 1.2.1. Figure 3-10 displays the mirror-to-source direction of the thought experiment for both scenarios. For each scenario, the midpoint reference frame is in a state of absolute rest.

The object is located in the midpoint reference frame, with the relative reference frame moving with speed $\frac{1}{2}v$ with respect to the midpoint reference frame, where v is the relative velocity between the event reference frame and the relative reference frame. The ruler within the midpoint reference frame still moves **away from** the relative reference frame. However, the light pulse travels with speed c **towards** the relative reference frame. In the time taken for the light pulse to travel from the mirror back to the light source, the relative reference frame has moved a distance $\frac{1}{2}v\Delta t_{r(2)}$ with respect to the midpoint reference frame. The distance travelled by the light pulse in the time interval $\Delta t_{r(2)}$ is given by:

$$d_{r(2)} = l_r - \frac{1}{2}v\Delta t_{r(2)} \qquad (3\text{-}24)$$

where $d_{r(2)}$ is the distance travelled by the light pulse in the time interval $\Delta t_{r(2)}$ as measured by the relative observer, l_r is the length of the ruler as measured by the relative observer, and v is the relative velocity between the event reference frame and the relative reference frame.

The light pulse travels with speed c, so equation (2-1) can be used to measure the distance $d_{r(2)}$ travelled by the light pulse in the time interval $\Delta t_{r(2)}$ to give:

$$d_{r(2)} = c\Delta t_{r(2)} \tag{3-25}$$

Equation (3-24) and equation (3-25) can be combined to eliminate $d_{r(2)}$ and solved in terms of $\Delta t_{r(2)}$ to get:

$$\Delta t_{r(2)} = \frac{l_r}{c + \frac{1}{2}v} \tag{3-26}$$

where $\Delta t_{r(2)}$ is the time taken for the light pulse to travel from the mirror back to the light source as measured by the relative observer, l_r is the length of the ruler as measured by the relative observer, v is the relative velocity between the event reference frame and the relative reference frame, and c is the speed of light.

The **total time** taken for the light pulse to travel from the light source to the mirror and then back to the light source is the sum of the time taken in the source-to-mirror direction $\Delta t_{r(1)}$ and the time taken in the mirror-to-source direction $\Delta t_{r(2)}$:

$$\Delta t_r = \frac{\left(c + \frac{1}{2}v\right)l_r + \left(c - \frac{1}{2}v\right)l_r}{c^2\left(1 - \frac{v^2}{4c^2}\right)} = \frac{2\gamma_m^2 l_r}{c} \tag{3-27}$$

where Δt_r is the time interval for the light pulse to travel from the light source to the mirror and then back to the light source as measured by the relative observer, and γ_m is the midpoint relativistic factor.

Equation (3-2) for Δt_m can be substituted into the midpoint-to-relative time dilation equation $\Delta t_t = \gamma_m \Delta t_m$ in (2-13) to give:

$$\Delta t_r = \frac{2\gamma_m l_m}{c} \tag{3-28}$$

Finally, equation (3-27) and equation (3-28) can be combined to eliminate Δt_r and solved for l_r to get:

$$\boxed{l_r = \frac{1}{\gamma_m} l_m} \tag{3-29}$$

where l_r is the length of the ruler as measured by the relative observer, γ_m is the midpoint relativistic factor, and l_m is the length of the ruler as measured by the midpoint observer.

The **inverse** value for the midpoint relativistic factor in equation (3-29) is always less than or equal to one, since the relative velocity is always less than twice the speed of light, due to the midpoint relative-velocity constraint. The **relative** observer measures a shorter length of the ruler compared to the **midpoint** observer's length measurement of the same ruler. This shortened length measurement by the relative observer is referred to as **midpoint-to-relative length contraction**.

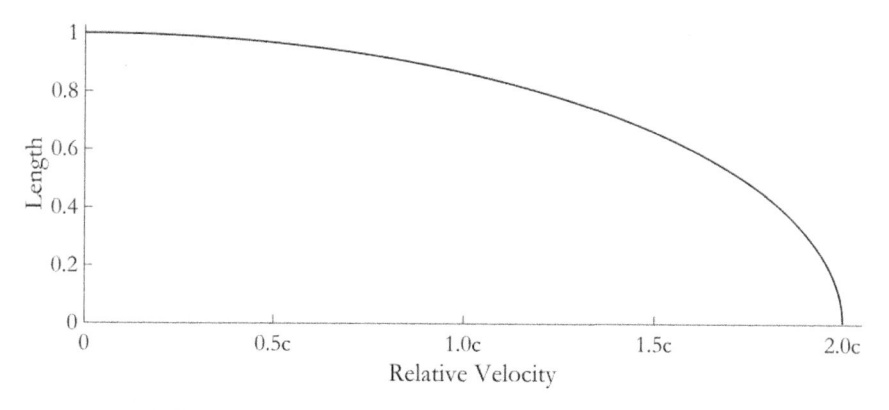

Figure 3-11. Midpoint-to-relative length contraction

Figure 3-11 displays the length-of-the-ruler values as measured by the relative observer against the relative-velocity values. To aid interpretation, all the length-of-the-ruler values are measured using equation (3-29) with a one-metre ruler in the midpoint reference frame ($l_m = 1$). The relative-velocity values are fractions of the speed of light. Table 3-3 reports a sample of the length-of-the-ruler values against the relative-velocity values.

Table 3-3. Midpoint-to-relative length contraction

l_m	Relative Velocity							
	0.00c	0.40c	0.80c	1.20c	1.60c	1.80c	1.90c	1.98c
1.00	1.00	0.98	0.92	0.80	0.60	0.44	0.31	0.14

The closer the relative velocity is to twice the speed of light, the larger the length contraction measured by the relative observer. For example, when the relative velocity between the event reference frame and the relative reference frame is 1.98c, a one-metre ruler within the midpoint reference frame is measured as 0.14 metres by the relative observer in the relative reference frame.

3.4.3 The complete solution

The **midpoint theory of relativity** divides the thought experiment into two parts: an event-to-midpoint part and a midpoint-to-relative part. The previous sections applied the special theory of relativity to each part. In this section, the measurements for the length of objects for both parts are combined to derive a complete measurement for the length of objects for the midpoint theory of relativity. Table 3-4 reports the formula used to measure the length of objects for both parts of midpoint relativity.

Table 3-4. Formula comparison for measuring the length of objects for both parts of midpoint relativity

Event-to-Midpoint	Midpoint-to-Relative
$l_m = \dfrac{1}{\gamma_m} l_e$	$l_r = \dfrac{1}{\gamma_m} l_m$

The event-to-midpoint part assumes that the midpoint reference frame is in a state of absolute rest and the event reference frame is in uniform motion. The length of an object formula for the midpoint observer is $l_m = l_e/\gamma_m$ from equation (3-20). The midpoint-to-relative part assumes that the midpoint reference frame is in a state of absolute rest and the relative reference frame is in uniform motion. The length of an object formula for the relative observer is $l_r = l_m/\gamma_m$ from equation (3-29).

The measurement formula for the length of an object for both parts of midpoint relativity can be combined by substituting equation (3-20) into equation (3-29) to obtain:

$$l_r = \frac{1}{\gamma_m^2} l_e \qquad (3\text{-}30)$$

where l_r is the length of the ruler as measured by the relative observer, γ_m is the midpoint relativistic factor, and l_e is the length of the ruler as measured by the event observer.

The **inverse** of the **squared** midpoint relativistic factor in equation (3-30) is always less than or equal to one, since the relative velocity is always less than twice the speed of light, due to the midpoint relative-velocity constraint. The **relative** observer measures a shorter length of an object compared to the **event** observer's length measurement of the same object. This decreased length measured by the relative observer is referred to as **midpoint length contraction**.

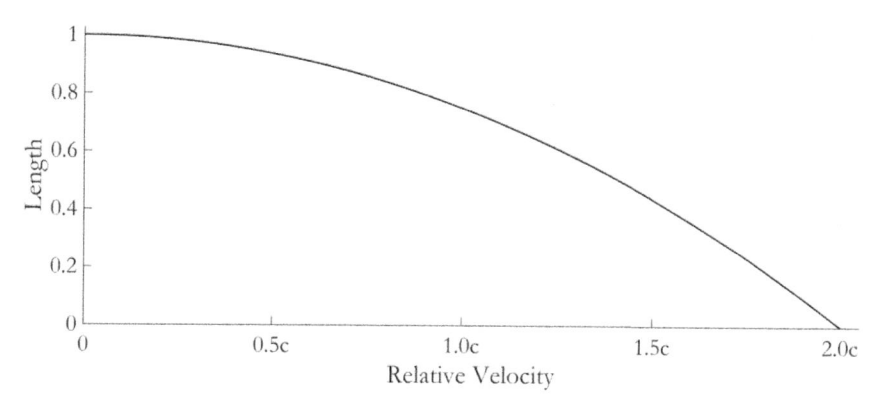

Figure 3-12. Midpoint length contraction

Figure 3-12 displays the length-of-the-ruler values as measured by the relative observer against the relative-velocity values. To aid interpretation, all the length-of-the-ruler values are measured using equation (3-30) with a one-metre ruler in the event reference frame ($l_e = 1$). The relative-velocity values are fractions of the speed of light. Table 3-5 reports a sample of the length-of-the-ruler values against the relative-velocity values.

Table 3-5. Midpoint length contraction

l_r	Relative Velocity							
	0.00c	0.40c	0.80c	1.20c	1.60c	1.80c	1.90c	1.98c
1.00	1.00	0.96	0.84	0.64	0.36	0.19	0.10	0.02

The closer the relative velocity is to the speed of light, the larger the length contraction measured by the relative observer. For example, when the relative velocity between the event reference frame and the relative reference frame is 1.98c, a one-metre ruler within the event reference frame is measured as 0.02 metres by the relative observer in the relative reference frame.

3.5 Relativity Comparison

This section compares the measurement formula for the length of objects for both special relativity and midpoint relativity. Table 3-6 reports the formula for the relative observer measuring the length of objects in the event reference frame.

Table 3-6. Formula comparison for measuring the length of objects

Special Relativity	Midpoint Relativity
$l_r = \dfrac{1}{\gamma_s} l_e$	$l_r = \dfrac{1}{\gamma_m^2} l_e$

The inverse relativistic factors drive the magnitude of the length contractions. The length of an object in the event reference frame is multiplied by the inverse of the squared midpoint relativistic factor in the midpoint theory of relativity, compared to the inverse of the single special relativistic factor in the special theory of relativity. This difference results in significantly different outcomes for both theories of relativity.

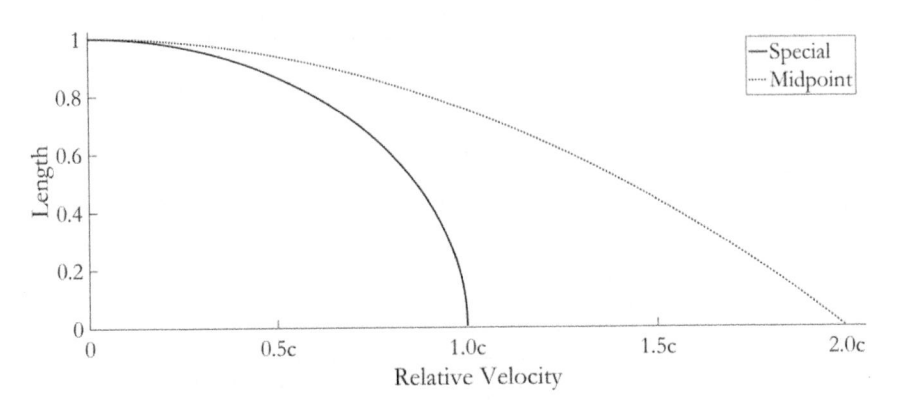

Figure 3-13. Length contraction comparison

Figure 3-13 displays the length-of-the-ruler values as measured by the relative observer against the relative-velocity values, for both special relativity and midpoint relativity. To aid interpretation, the length-of-the-ruler values are measured using equation (3-11) for special relativity and equation (3-30) for midpoint relativity, with a one-metre ruler in the event reference frame ($l_e = 1$). The relative-velocity values are fractions of the speed of light. Table 3-7 reports a sample of the length-of-the-ruler values against the relative-velocity values for both special relativity (SR) and midpoint relativity (MR).

Table 3-7. Length contraction comparison

	l_e	\multicolumn{8}{c}{Relative Velocity}
		0.00c	0.50c	0.90c	0.95c	0.99c	1.80c	1.90c	1.98c
SR	1.00	1.00	0.87	0.44	0.31	0.14			
MR	1.00	1.00	0.94	0.80	0.77	0.76	0.19	0.10	0.02

The special theory of relativity is unable to measure the length of objects when the relative velocity is greater than, or equal to, the speed of light, due to the special relative-velocity constraint. In contrast, the midpoint theory of relativity has a relative-velocity constraint of less than **twice** the speed of light, due to the midpoint relative-velocity constraint.

When the relative velocity is less than the speed of light, the special theory of relativity overestimates length contraction compared to the midpoint theory of relativity, with most of the overestimation occurring when the relative velocity is close to the speed of light. For example, when the relative velocity between the event reference frame and the relative reference frame is $0.99c$, a one-metre ruler in the event reference frame is measured by the relative observer as 0.14 metres using the special theory of relativity, compared to 0.76 metres using the midpoint theory of relativity.

4. Frequency Shift

The measurement of the **frequency shift** in electromagnetic waves is dependent on the observer's reference frame. If a light source is in the observer's reference frame, all the normal laws of physics apply. In contrast, if a light source is in another reference frame, the laws of relativity apply.

4.1 Thought Experiment

A thought experiment is used to demonstrate the measurement of the frequency shift in electromagnetic waves from another reference frame. The thought experiment consists of a light source that emits light pulses with constant frequency.

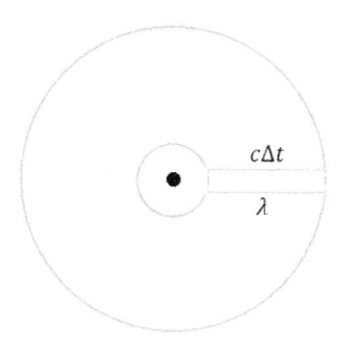

Figure 4-1. Thought experiment to measure the frequency shift

An observer measures the period (time interval) of light pulses. Figure 4-1 displays the thought experiment, where the centre dot represents the light source and the outer circles represent wave crests. Using the normal laws of physics, the period of the light pulse is given by:

$$\Delta t = \frac{\lambda}{c} = \frac{1}{f} \tag{4-1}$$

where Δt is the period (time interval) of the light source, λ is the wavelength of the light source, $f = c/\lambda$ is the frequency of the light source, and c is the speed of light. The thought experiment is used to measure the frequency shift in electromagnetic waves from the perspective of both special relativity and midpoint relativity.

4.2 Principle of Relativity

The principle of relativity states that all the normal laws of physics are the same in all reference frames. Both special relativity and midpoint relativity use observers in their respective thought experiments.

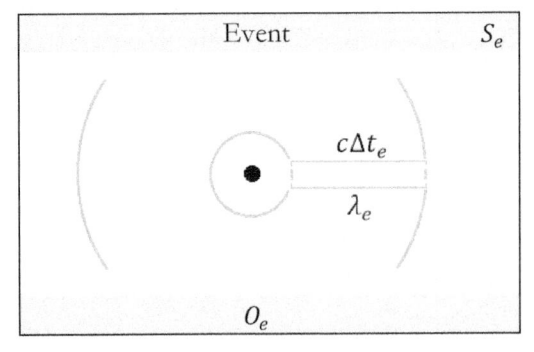

Figure 4-2. Frequency measurement for the event observer

The light source of the thought experiment is located in the event reference frame, which contains the event observer. Figure 4-2 displays the thought experiment for the event observer O_e in the event reference frame S_e. All the normal laws of physics apply because the light source is located in the event observer's reference frame. Thus, equation (4-1) can be used to measure the period (time interval) of the light pulse:

$$\Delta t_e = \frac{\lambda_e}{c} = \frac{1}{f_e} \qquad (4\text{-}2)$$

where Δt_e is the period of the light source as measured by the event observer, λ_e is the wavelength of the light source as measured by the event observer, $f_e = c/\lambda_e$ is the frequency of the light source as measured by the event observer, and c is the speed of light.

4.3 Special Relativity

The special theory of relativity is used to measure the **frequency shift** for a light source emitting light pulses in another reference frame from the observer. More specifically, the light source is located in the event reference frame and is measured by the relative observer in the relative reference frame. The special theory of relativity assumes that one of the reference frames is in a state of absolute rest.

The **two directions** of travel of the light source are important for measuring the frequency shift in electromagnetic waves. The first direction is the light source moving **towards** the receiver. The second direction is the light source moving **away from** the receiver.

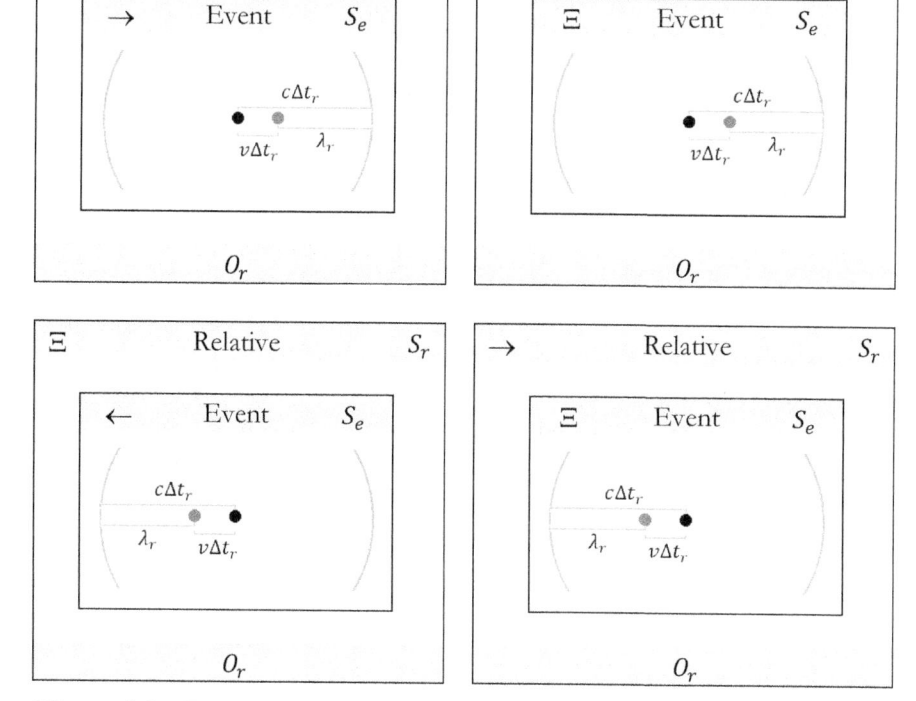

Figure 4-3. Frequency shift for the light source moving towards the receiver for the four scenarios of special relativity

For the light source moving **towards** the receiver, the special theory of relativity has four different scenarios for the two reference frames, which are discussed in Section 1.1.1. Figure 4-3 displays the thought experiment for all four scenarios. For each scenario, one reference frame is assumed to be in a state of absolute rest.

The light source is located in the event reference frame. By the time the light source emits the next light pulse, the light source (event reference frame) has moved a distance $v\Delta t_r$ towards the receiver (relative reference frame) for all four scenarios. The wavelength of the light pulse is given by:

$$\lambda_r = (c - v)\Delta t_r \qquad (4\text{-}3)$$

and the frequency of the light pulse is given by:

$$f_r = \frac{c}{\lambda_r} = \frac{1}{\left(1 - \frac{v}{c}\right)\Delta t_r} \qquad (4\text{-}4)$$

where λ_r is the wavelength of the light pulse as measured by the relative observer, f_r is the frequency of the light pulse as measured by the relative observer, Δt_r is the period between successive wave crests as measured by the relative observer, v is the relative velocity between the event reference frame and the relative reference frame, and c is the speed of light.

Equation (4-4) can be rearranged in terms of Δt_r to give:

$$\Delta t_r = \frac{1}{\left(1 - \frac{v}{c}\right) f_r} \qquad (4\text{-}5)$$

The special time dilation relationship between Δt_r and Δt_e in equation (2-5) can be used to produce:

$$\Delta t_r = \frac{\Delta t_e}{\sqrt{1 - \frac{v^2}{c^2}}} = \frac{1}{f_e \sqrt{1 - \frac{v^2}{c^2}}} \qquad (4\text{-}6)$$

where $\Delta t_e = 1/f_e$ from equation (4-2). Equation (4-5) and equation (4-6) can be combined to eliminate Δt_r to obtain:

$$\left(1 - \frac{v}{c}\right) f_r = f_e \sqrt{1 - \frac{v^2}{c^2}} = f_e \sqrt{\left(1 + \frac{v}{c}\right)\left(1 - \frac{v}{c}\right)} \qquad (4\text{-}7)$$

Finally, equation (4-7) can be rewritten in terms of f_r to give:

$$f_r = f_e \frac{\sqrt{c+v}}{\sqrt{c-v}} \qquad (4\text{-}8)$$

where f_r is the frequency of the light pulse as measured by the relative observer, f_e is the frequency of the light pulse as measured by the event observer, v is the relative velocity between the event reference frame and the relative reference frame, and c is the speed of light.

The value for $\sqrt{c+v}/\sqrt{c-v}$ in equation (4-8) is always greater than or equal to one, since the relative velocity is always less than the speed of light, due to the special relative-velocity constraint. The relative observer measures a higher frequency, and a shorter wavelength, when the **event** reference frame moves **towards** the **relative** reference frame, which is referred to as the **special blueshift Doppler effect**.

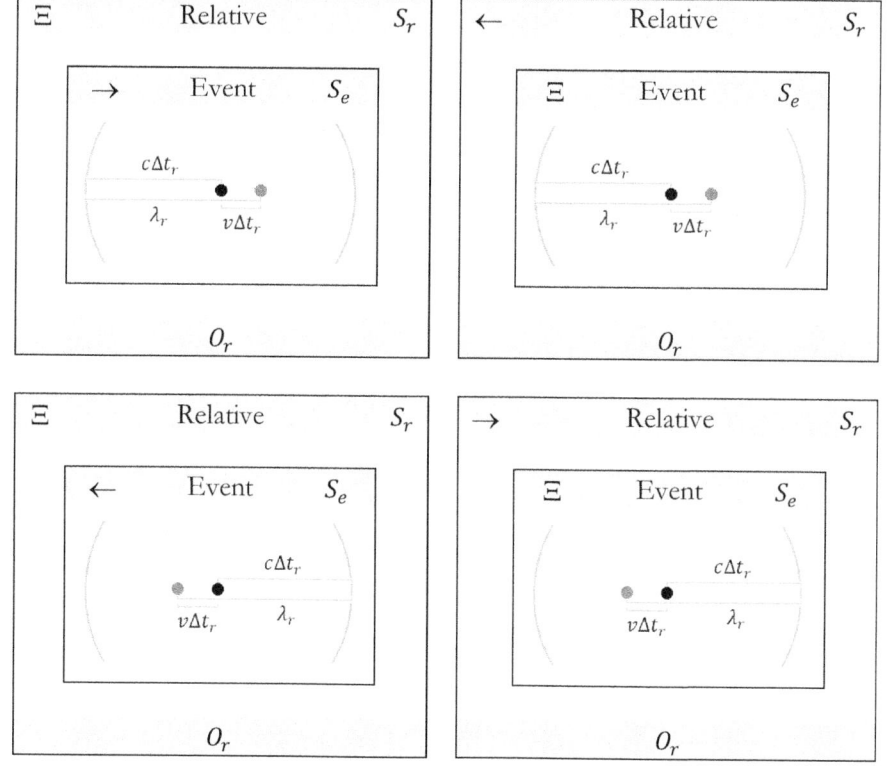

Figure 4-4. Frequency shift for the light source moving away from the receiver for the four scenarios of special relativity

For the light source moving **away from** the receiver, the special theory of relativity has four different scenarios for the two reference frames, which are discussed in Section 1.1.1. Figure 4-4 displays the thought experiment for all four scenarios. For each scenario, one reference frame is assumed to be in a state of absolute rest.

The light source is located in the event reference frame. By the time the light source emits the next light pulse, the light source (event reference frame) has moved a distance $v\Delta t_r$ away from the receiver (relative reference frame) for all four scenarios. The wavelength of the light pulse is given by:

$$\lambda_r = (c + v)\Delta t_r \tag{4-9}$$

and the frequency of the light pulse is given by:

$$f_r = \frac{c}{\lambda_r} = \frac{1}{\left(1 + \frac{v}{c}\right)\Delta t_r} \tag{4-10}$$

where λ_r is the wavelength of the light pulse as measured by the relative observer, f_r is the frequency of the light pulse as measured by the relative observer, Δt_r is the period between successive wave crests as measured by the relative observer, v is the relative velocity between the event reference frame and the relative reference frame, and c is the speed of light.

Equation (4-10) can be rearranged in terms of Δt_r to give:

$$\Delta t_r = \frac{1}{\left(1 + \frac{v}{c}\right)f_r} \tag{4-11}$$

The special time dilation relationship between Δt_r and Δt_e in equation (2-5) can be used to produce:

$$\Delta t_r = \frac{\Delta t_e}{\sqrt{1 - \frac{v^2}{c^2}}} = \frac{1}{f_e\sqrt{1 - \frac{v^2}{c^2}}} \tag{4-12}$$

where $\Delta t_e = 1/f_e$ from equation (4-2). Equation (4-11) and equation (4-12) can be combined to eliminate Δt_r to obtain:

$$\left(1 + \frac{v}{c}\right)f_r = f_e\sqrt{1 - \frac{v^2}{c^2}} = f_e\sqrt{\left(1 + \frac{v}{c}\right)\left(1 - \frac{v}{c}\right)} \tag{4-13}$$

Finally, equation (4-13) can be rewritten in terms of f_r to give:

$$f_r = f_e \frac{\sqrt{c-v}}{\sqrt{c+v}} \qquad (4\text{-}14)$$

where f_r is the frequency of the light pulse as measured by the relative observer, f_e is the frequency of the light pulse as measured by the event observer, v is the relative velocity between the event reference frame and the relative reference frame, and c is the speed of light.

The value for $\sqrt{c-v}/\sqrt{c+v}$ in equation (4-14) is always less than or equal to one, since the relative velocity is always less than the speed of light, due to the special relative-velocity constraint. The relative observer measures a lower frequency, and longer wavelength, when the **event** reference frame moves **away from** the **relative** reference frame, which is referred to as the **special red-shift Doppler effect**.

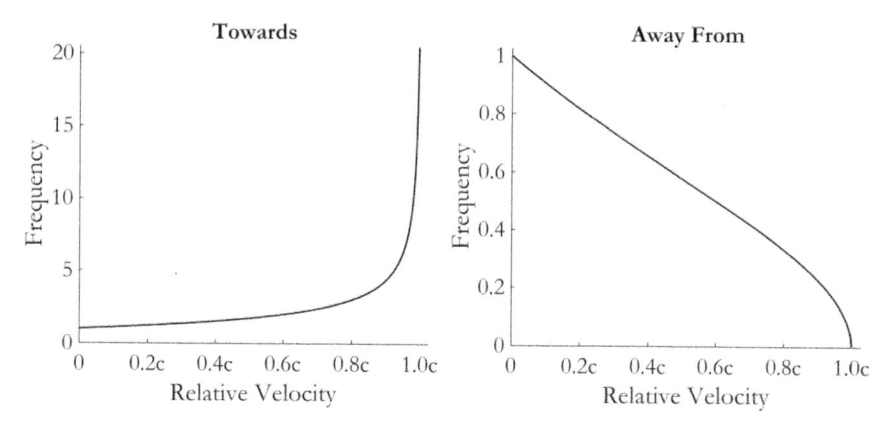

Figure 4-5. Special Doppler effect

Figure 4-5 displays the frequency-shift values as measured by the relative observer against the relative-velocity values. To aid interpretation, all frequency-shift values are measured with a frequency of one ($f_e = 1$) in the event reference frame against different relative-velocity values, which are fractions of the speed of light. Table 4-1 reports a sample of the frequency-shift values against the relative-velocity values.

Table 4-1. Special Doppler effect

Direction	Relative Velocity							
	0.00c	0.20c	0.40c	0.60c	0.80c	0.90c	0.95c	0.99c
Towards	1.00	1.22	1.53	2.00	3.00	4.36	6.24	14.11
Away From	1.00	0.82	0.65	0.50	0.33	0.23	0.16	0.07

The closer the relative velocity is to the speed of light, the larger the frequency shift in electromagnetic waves measured by the relative observer. For example, when the event reference frame is moving **towards** the relative reference frame with a relative velocity of 0.99c, a frequency of one ($f_e = 1$) in the event reference frame is measured as a frequency of 14.11 by the relative observer. In contrast, when the event reference frame is moving **away from** the relative reference frame with a relative velocity of 0.99c, a frequency of one ($f_e = 1$) in the event reference frame is measured as a frequency of 0.07 by the relative observer.

4.4 Midpoint Relativity

The midpoint theory of relativity does **not** assume that the event reference frame or the relative reference frame is in a state of absolute rest. In a system that consists of two **standard** reference frames in uniform motion, the midpoint is the only point that is in a state of absolute rest. The midpoint is the equidistant point between the light source in the event reference frame and the relative observer in the relative reference frame. Midpoint relativity creates a **phantom** midpoint reference frame with a phantom midpoint observer. The midpoint reference frame is always in a state of absolute of rest.

The thought experiment used to demonstrate the measurement of the frequency shift in electromagnetic waves for the special theory of relativity, is reused to measure the frequency shift for the midpoint theory of relativity. The thought experiment consists of a light source that emits light pulses with frequency equal to the frequency of the light pulse. To understand the measurement of the frequency shift in electromagnetic waves using the midpoint theory of relativity, the thought experiment can be divided into two parts: an **event-to-midpoint part** and a **midpoint-to-relative part**. The special theory of relativity can be applied separately to both parts. The measurement of the frequency shift in electromagnetic waves for each part can be combined to form a complete measurement of the frequency shift in electromagnetic waves for the midpoint theory of relativity.

4.4.1   The event-to-midpoint part

The event-to-midpoint part assumes that the midpoint reference frame is in a state of absolute rest and the event reference frame is travelling with uniform motion either to the left or to the right. The **two directions** of travel of the light source are important for measuring the frequency shift in electromagnetic waves. The first direction is the light source moving **towards** the receiver. The second direction is the light source moving **away from** the receiver.

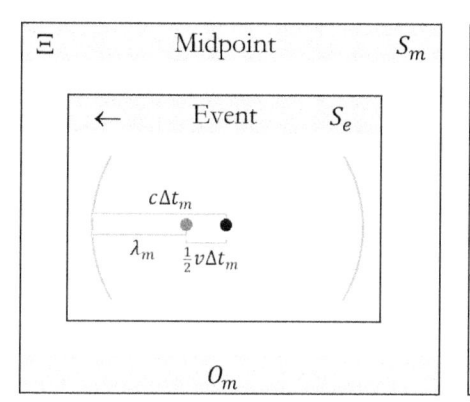

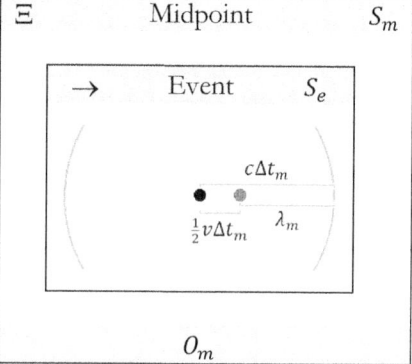

Figure 4-6. Frequency shift for the light source moving towards the receiver for the two scenarios of the event-to-midpoint part

For the light source moving **towards** the receiver, the event-to-midpoint part of the midpoint theory of relativity has two different scenarios for the two reference frames, which are discussed in Section 1.2.1. Figure 4-6 displays the thought experiment for both scenarios. For each scenario, the midpoint reference frame is in a state of absolute rest.

The light source is located in the event reference frame. By the time the light source emits the next light pulse, the event reference frame has moved a distance $\frac{1}{2}v\Delta t_m$ towards the midpoint reference frame. The wavelength of the light pulse is given by:

$$\lambda_m = \left(c - \frac{v}{2}\right)\Delta t_m \qquad (4\text{-}15)$$

and the frequency of the light pulse is given by:

$$f_m = \frac{c}{\lambda_m} = \frac{1}{\left(1 - \frac{v}{2c}\right)\Delta t_m} \qquad (4\text{-}16)$$

where λ_m is the wavelength of the light pulse as measured by the midpoint observer, f_m is the frequency of the light pulse as measured by the midpoint observer, Δt_m is the period between successive wave crests as measured by the midpoint observer, v is the relative velocity between the event reference frame and the relative reference frame, and c is the speed of light.

Equation (4-16) can be rearranged in terms of Δt_m to give:

$$\Delta t_m = \frac{1}{\left(1 - \frac{v}{2c}\right) f_m} \tag{4-17}$$

The event-to-midpoint time dilation relationship between Δt_m and Δt_e in equation (2-9) can be used to produce:

$$\Delta t_m = \frac{\Delta t_e}{\sqrt{1 - \frac{v^2}{4c^2}}} = \frac{1}{f_e \sqrt{1 - \frac{v^2}{4c^2}}} \tag{4-18}$$

where $\Delta t_e = 1/f_e$ from equation (4-2). Equation (4-17) and equation (4-18) can be combined to eliminate Δt_m to obtain:

$$\left(1 - \frac{v}{2c}\right) f_m = f_e \sqrt{1 - \frac{v^2}{4c^2}} = f_e \sqrt{\left(1 + \frac{v}{2c}\right)\left(1 - \frac{v}{2c}\right)} \tag{4-19}$$

Finally, equation (4-19) can be rewritten in terms of f_m to give:

$$\boxed{f_m = f_e \frac{\sqrt{2c + v}}{\sqrt{2c - v}}} \tag{4-20}$$

where f_m is the frequency of the light pulse as measured by the midpoint observer, f_e is the frequency of the light pulse as measured by the event observer, v is the relative velocity between the event reference frame and the relative reference frame, and c is the speed of light.

The value for $\sqrt{2c + v}/\sqrt{2c - v}$ in equation (4-20) is always greater than or equal to one, since the relative velocity is always less than **twice** the speed of light, due to the midpoint relative-velocity constraint. The midpoint observer measures a higher frequency, and a shorter wavelength, when the **event** reference frame is moving **towards** the **midpoint** reference frame, which is referred to as the **event-to-midpoint blue-shift Doppler effect**.

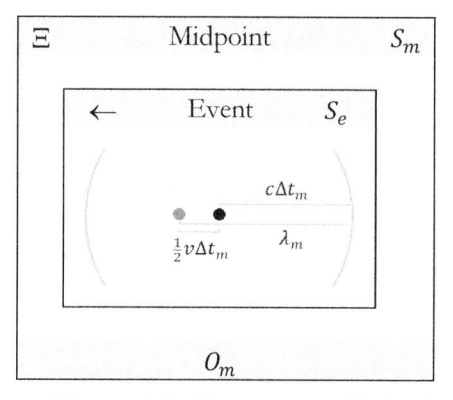

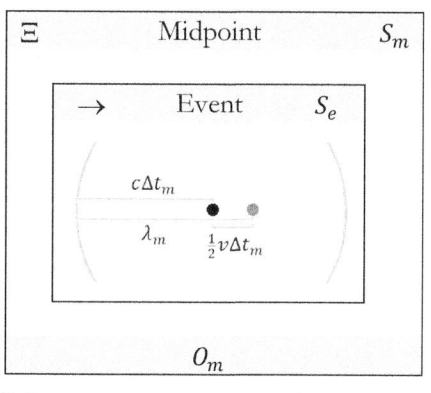

Figure 4-7. Frequency shift for the light source moving away from the receiver for the two scenarios of the event-to-midpoint part

For the light source moving **away from** the receiver, the event-to-midpoint part of the midpoint theory of relativity has two different scenarios for the two reference frames, which are discussed in Section 1.2.1. Figure 4-7 displays the thought experiment for both scenarios. For each scenario, the midpoint reference frame is in a state of absolute rest.

The light source is located in the event reference frame. By the time the light source emits the next light pulse, the event reference frame has moved a distance $\frac{1}{2}v\Delta t_m$ away from the midpoint reference frame. The wavelength of the light pulse is given by:

$$\lambda_m = \left(c + \frac{v}{2}\right)\Delta t_m \qquad (4\text{-}21)$$

and the frequency of the light pulse is given by:

$$f_m = \frac{c}{\lambda_m} = \frac{1}{\left(1 + \frac{v}{2c}\right)\Delta t_m} \qquad (4\text{-}22)$$

where λ_m is the wavelength of the light pulse as measured by the midpoint observer, f_m is the frequency of the light pulse as measured by the midpoint observer, Δt_m is the period between successive wave crests as measured by the midpoint observer, v is the relative velocity between the event reference frame and the relative reference frame, and c is the speed of light.

Equation (4-22) can be rearranged in terms of Δt_m to give:

$$\Delta t_m = \frac{1}{\left(1 + \frac{v}{2c}\right) f_m} \qquad (4\text{-}23)$$

The event-to-midpoint time dilation relationship between Δt_m and Δt_e in equation (2-9) can be used to produce:

$$\Delta t_m = \frac{\Delta t_e}{\sqrt{1 - \frac{v^2}{4c^2}}} = \frac{1}{f_e \sqrt{1 - \frac{v^2}{4c^2}}} \qquad (4\text{-}24)$$

where $\Delta t_e = 1/f_e$ from equation (4-2). Equation (4-23) and equation (4-24) can be combined to eliminate Δt_m to obtain:

$$\left(1 + \frac{v}{2c}\right) f_m = f_e \sqrt{1 - \frac{v^2}{4c^2}} = f_e \sqrt{\left(1 + \frac{v}{2c}\right)\left(1 - \frac{v}{2c}\right)} \qquad (4\text{-}25)$$

Finally, equation (4-25) can be rewritten in terms of f_m to give:

$$\boxed{f_m = f_e \frac{\sqrt{2c - v}}{\sqrt{2c + v}}} \qquad (4\text{-}26)$$

where f_m is the frequency of the light pulse as measured by the midpoint observer, f_e is the frequency of the light pulse as measured by the event observer, v is the relative velocity between the event reference frame and the relative reference frame, and c is the speed of light.

The value for $\sqrt{2c - v}/\sqrt{2c + v}$ in equation (4-26) is always less than or equal to one, since the relative velocity is always less than **twice** the speed of light, due to the midpoint relative-velocity constraint. The midpoint observer measures a lower frequency, and longer wavelength, when the **event** reference frame moves **away from** the **midpoint** reference frame, which is referred to as the **event-to-midpoint red-shift Doppler effect**.

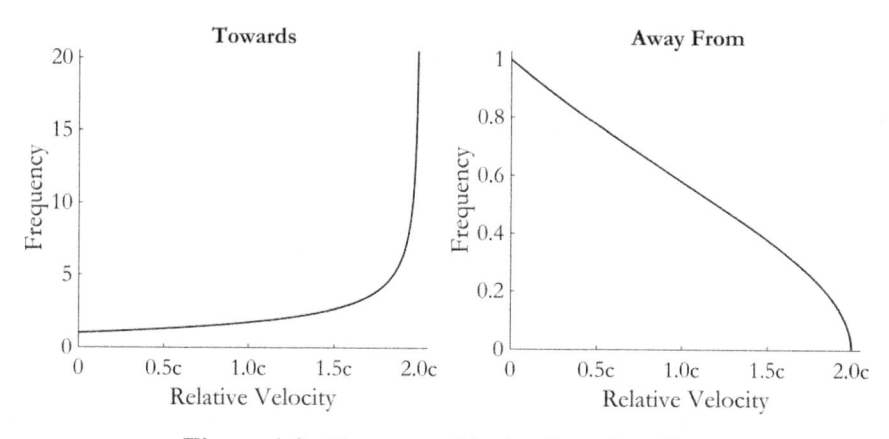

Figure 4-8. Event-to-midpoint Doppler effect

Figure 4-8 displays the frequency-shift values as measured by the midpoint observer against the relative-velocity values. To aid interpretation, all frequency-shift values are measured with a frequency of one ($f_e = 1$) in the event reference frame against different relative-velocity values, which are fractions of the speed of light. Table 4-2 reports a sample of the frequency-shift values against the relative-velocity values.

Table 4-2. Event-to-midpoint Doppler effect

Direction	\multicolumn{8}{c}{Relative Velocity}
	0.00c	0.40c	0.80c	1.20c	1.60c	1.80c	1.90c	1.98c
Towards	1.00	1.22	1.53	2.00	3.00	4.36	6.24	14.11
Away From	1.00	0.82	0.65	0.50	0.33	0.23	0.16	0.07

The closer the relative velocity is to twice the speed of light, the larger the frequency shift in electromagnetic waves measured by the midpoint observer, where the relative velocity is between the event reference frame and the relative reference frame. For example, when the relative velocity is 1.98c with the event reference frame moving **towards** the midpoint reference frame, a frequency of one ($f_e = 1$) in the event reference frame is measured as a frequency of 14.11 by the midpoint observer. In contrast, when the relative velocity is 1.98c with the event reference frame moving **away from** the midpoint reference frame, a frequency of one ($f_e = 1$) in the event reference frame is measured as a frequency of 0.07 by the midpoint observer.

4.4.2 The midpoint-to-relative part

The midpoint-to-relative part assumes that the midpoint reference frame is in a state of absolute rest and the relative reference frame is travelling with uniform motion either to the left or to the right. The **two directions** of travel of the light source are important for measuring the frequency shift in electromagnetic waves. The first direction is the light source moving **towards** the receiver. The second direction is the light source moving **away from** the receiver.

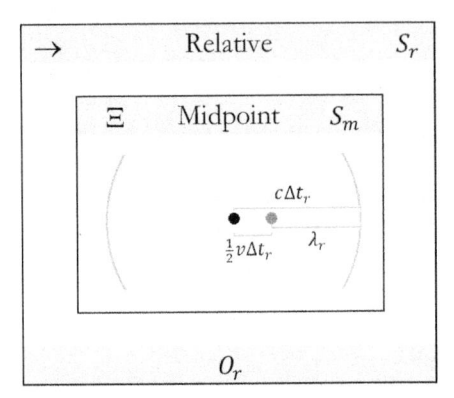

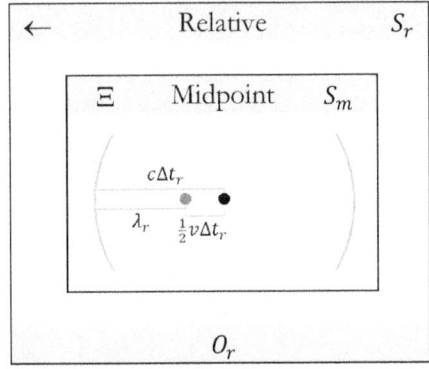

Figure 4-9. Frequency shift for the light source moving towards the receiver for the two scenarios of the midpoint-to-relative part

For the light source moving **towards** the receiver, the midpoint-to-relative part of the midpoint theory of relativity has two different scenarios for the two reference frames, which are discussed in Section 1.2.1. Figure 4-9 displays the thought experiment for both scenarios. For each scenario, the midpoint reference frame is in a state of absolute rest.

The light source is located in the midpoint reference frame. By the time the light source emits the next light pulse, the relative reference frame has moved a distance $\frac{1}{2}v\Delta t_m$ towards the midpoint reference frame. The wavelength of the light pulse is given by:

$$\lambda_r = \left(c - \tfrac{1}{2}v\right)\Delta t_r \qquad (4\text{-}27)$$

and the frequency of the light pulse is given by:

$$f_r = \frac{c}{\lambda_r} = \frac{1}{\left(1 - \frac{v}{2c}\right)\Delta t_r} \qquad (4\text{-}28)$$

where λ_r is the wavelength of the light pulse as measured by the relative observer, f_r is the frequency of the light pulse as measured by the relative observer, Δt_r is the period between successive wave crests as measured by the relative observer, v is the relative velocity between the event reference frame and the relative reference frame, and c is the speed of light.

Equation (4-28) can be rearranged in terms of Δt_r to give:

$$\Delta t_r = \frac{1}{\left(1 - \frac{v}{2c}\right) f_r} \qquad (4\text{-}29)$$

The midpoint-to-relative time dilation relationship between Δt_r and Δt_m in equation (2-13) can be used to produce:

$$\Delta t_r = \frac{\Delta t_m}{\sqrt{1 - \frac{v^2}{4c^2}}} = \frac{1}{f_m \sqrt{1 - \frac{v^2}{4c^2}}} \qquad (4\text{-}30)$$

where $\Delta t_m = 1/f_m$ from equation (4-2). Equation (4-29) and equation (4-30) can be combined to eliminate Δt_r to obtain:

$$\left(1 - \frac{v}{2c}\right) f_r = f_m \sqrt{1 - \frac{v^2}{4c^2}} = f_m \sqrt{\left(1 + \frac{v}{2c}\right)\left(1 - \frac{v}{2c}\right)} \qquad (4\text{-}31)$$

Finally, equation (4-31) can be rewritten in terms of f_r to give:

$$\boxed{f_r = f_m \frac{\sqrt{2c + v}}{\sqrt{2c - v}}} \qquad (4\text{-}32)$$

where f_r is the frequency of the light pulse as measured by the relative observer, f_m is the frequency of the light pulse as measured by the midpoint observer, v is the relative velocity between the event reference frame and the relative reference frame, and c is the speed of light.

The value for $\sqrt{2c + v}/\sqrt{2c - v}$ in equation (4-32) is always greater than or equal to one, since the relative velocity is always less than **twice** the speed of light, due to the midpoint relative-velocity constraint. The relative observer measures a higher frequency, and a shorter wavelength, when the **relative** reference frame moves **towards** the **midpoint** reference frame, which is referred to as the **midpoint-to-relative blue-shift Doppler effect**.

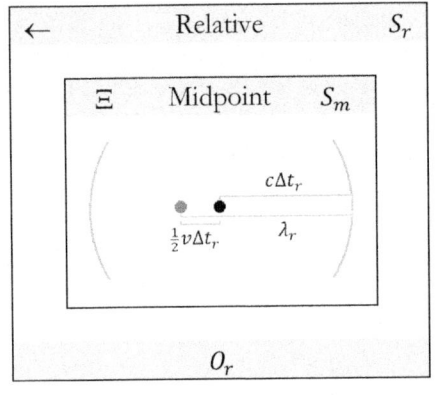

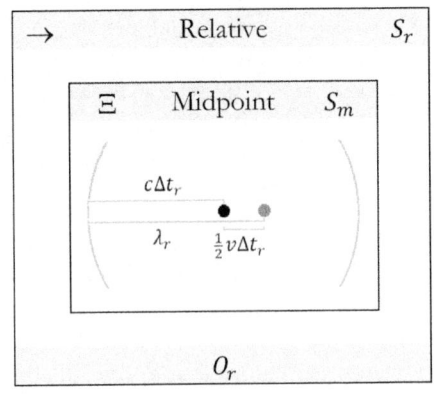

Figure 4-10. Frequency shift for the light source moving away from the receiver for the two scenarios of the midpoint-to-relative part

For the light source moving **away from** the receiver, the midpoint-to-relative part of the midpoint theory of relativity has two different scenarios for the two reference frames, which are discussed in Section 1.2.1. Figure 4-10 displays the thought experiment for both scenarios. For each scenario, the midpoint reference frame is in a state of absolute rest.

The light source is located in the midpoint reference frame. By the time the light source emits the next light pulse, the relative reference frame has moved a distance $\frac{1}{2}v\Delta t_m$ away from the midpoint reference frame. The wavelength of the light pulse is given by:

$$\lambda_r = \left(c + \frac{1}{2}v\right)\Delta t_r \qquad (4\text{-}33)$$

and the frequency of the light pulse is given by:

$$f_r = \frac{c}{\lambda_r} = \frac{1}{\left(1 + \frac{v}{2c}\right)\Delta t_r} \qquad (4\text{-}34)$$

where λ_r is the wavelength of the light pulse as measured by the relative observer, f_r is the frequency of the light pulse as measured by the relative observer, Δt_r is the period between successive wave crests as measured by the relative observer, v is the relative velocity between the event reference frame and the relative reference frame, and c is the speed of light.

Equation (4-34) can be rearranged in terms of Δt_r to give:

$$\Delta t_r = \frac{1}{\left(1 + \frac{v}{2c}\right) f_r} \qquad (4\text{-}35)$$

The midpoint-to-relative time dilation relationship between Δt_r and Δt_m in equation (2-13) can be used to produce:

$$\Delta t_r = \frac{\Delta t_m}{\sqrt{1 - \frac{v^2}{4c^2}}} = \frac{1}{f_m \sqrt{1 - \frac{v^2}{4c^2}}} \qquad (4\text{-}36)$$

where $\Delta t_m = 1/f_m$ from equation (4-2). Equation (4-35) and equation (4-36) can be combined to eliminate Δt_r to obtain:

$$\left(1 + \frac{v}{2c}\right) f_r = f_m \sqrt{1 - \frac{v^2}{4c^2}} = f_m \sqrt{\left(1 + \frac{v}{2c}\right)\left(1 - \frac{v}{2c}\right)} \qquad (4\text{-}37)$$

Finally, equation (4-37) can be rewritten in terms of f_r to give:

$$\boxed{f_r = f_m \frac{\sqrt{2c - v}}{\sqrt{2c + v}}} \qquad (4\text{-}38)$$

where f_r is the frequency of the light pulse as measured by the relative observer, f_m is the frequency of the light pulse as measured by the midpoint observer, v is the relative velocity between the event reference frame and the relative reference frame, and c is the speed of light.

The value for $\sqrt{2c - v}/\sqrt{2c + v}$ in equation (4-38) is always less than or equal to one, since the relative velocity is always less than **twice** the speed of light, due to the midpoint relative-velocity constraint. The relative observer measures a lower frequency, and a longer wavelength, when the **relative** reference frame moves **away from** the **midpoint** reference frame, which is referred to as the **midpoint-to-relative red-shift Doppler effect**.

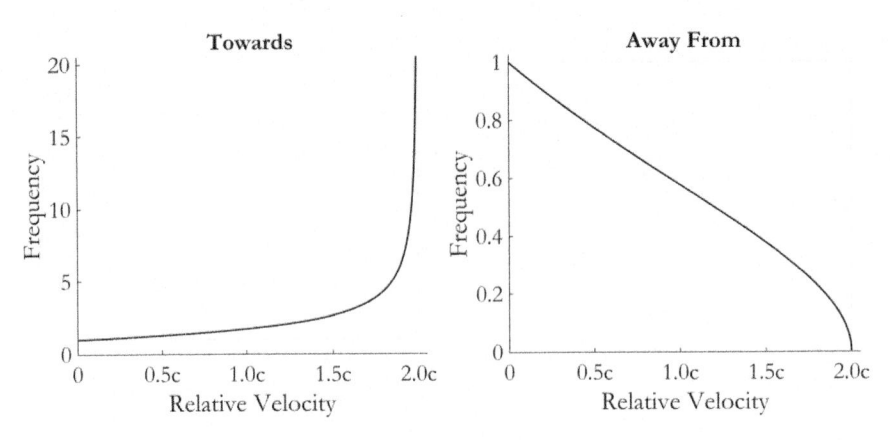

Figure 4-11. Midpoint-to-relative Doppler effect

Figure 4-11 displays the frequency-shift values as measured by the relative observer against the relative-velocity values. To aid interpretation, all frequency-shift values are measured with a frequency of one ($f_m = 1$) in the midpoint reference frame against different relative-velocity values, which are fractions of the speed of light. Table 4-3 reports a sample of the frequency-shift values against the relative-velocity values.

Table 4-3. Midpoint-to-relative Doppler effect

Direction	\multicolumn{8}{c}{Relative Velocity}
	0.00c	0.40c	0.80c	1.20c	1.60c	1.80c	1.90c	1.98c
Towards	1.00	1.22	1.53	2.00	3.00	4.36	6.24	14.11
Away From	1.00	0.82	0.65	0.50	0.33	0.23	0.16	0.07

The closer the relative velocity is to twice the speed of light, the larger the frequency shift in electromagnetic waves measured by the relative observer, where the relative velocity is between the event reference frame and the relative reference frame. For example, when the relative velocity is 1.98c with the relative reference frame moving **towards** the midpoint reference frame, a frequency of one ($f_m = 1$) in the midpoint reference frame is measured as a frequency of 14.11 by the relative observer. In contrast, when the relative velocity is 1.98c with the relative reference frame moving **away from** the midpoint reference frame, a frequency of one ($f_m = 1$) in the midpoint reference frame is measured as a frequency of 0.07 by the relative observer.

4.4.3 The complete solution

The **midpoint theory of relativity** divides the thought experiment into two parts: an event-to-midpoint part and a midpoint-to-relative part. The previous sections applied the special theory of relativity to each part. In this section, the measurements for the frequency shift in electromagnetic waves for both parts are combined to derive a complete measure of the frequency shift in electromagnetic waves for the midpoint theory of relativity. Table 4-4 reports the formula for measuring the frequency shift in electromagnetic waves for both parts of midpoint relativity.

Table 4-4. Formula comparison for measuring the frequency shift for both parts of midpoint relativity

	Event-to-Midpoint	Midpoint-to-Relative
Towards	$f_m = f_e \dfrac{\sqrt{2c+v}}{\sqrt{2c-v}}$	$f_r = f_m \dfrac{\sqrt{2c+v}}{\sqrt{2c-v}}$
Away From	$f_m = f_e \dfrac{\sqrt{2c-v}}{\sqrt{2c+v}}$	$f_r = f_m \dfrac{\sqrt{2c-v}}{\sqrt{2c+v}}$

The **two directions** of travel of the light source are important for measuring the frequency shift in electromagnetic waves. The results for the light source moving **towards** the receiver are presented first, which are followed by the results for the light source moving **away from** the receiver.

For the light source moving **towards** the receiver, the event-to-midpoint part assumes that the midpoint reference frame is in a state of absolute rest and the event reference frame is in uniform motion. The formula for measuring the frequency shift in electromagnetic waves for the midpoint observer is $f_m = f_e\sqrt{2c+v}/\sqrt{2c-v}$ from equation (4-20). The midpoint-to-relative part assumes that the midpoint reference frame is in a state of absolute rest and the relative reference frame is in uniform motion. The formula for measuring the frequency shift in electromagnetic waves for the relative observer is $f_r = f_m\sqrt{2c+v}/\sqrt{2c-v}$ from equation (4-32). The measurements of the frequency shift in electromagnetic waves for both parts can be combined by substituting equation (4-20) into equation (4-32) to obtain:

$$f_r = f_e \left(\frac{2c+v}{2c-v}\right) \quad (4\text{-}39)$$

where f_r is the frequency of the light pulse as measured by the relative observer, f_e is the frequency of the light pulse as measured by the event observer, v is the relative velocity between the event reference frame and the relative reference frame, and c is the speed of light.

The value for $(2c + v)/(2c - v)$ in equation (4-39) is always greater than or equal to one, since the relative velocity is always less than **twice** the speed of light, due to the midpoint relative-velocity constraint. The relative observer measures a higher frequency, and a shorter wavelength, when the **event** reference frame moves **towards** the **relative** reference frame, which is referred to as the **midpoint blue-shift Doppler effect**.

For the light source moving **away from** the receiver, the event-to-midpoint part assumes that the midpoint reference frame is in a state of absolute rest and the event reference frame is in uniform motion. The formula for measuring the frequency shift in electromagnetic waves for the midpoint observer is $f_m = f_e \sqrt{2c - v}/\sqrt{2c + v}$ from equation (4-26). The midpoint-to-relative part assumes that the midpoint reference frame is in a state of absolute rest and the relative reference frame is in uniform motion. The formula for measuring the frequency shift in electromagnetic waves for the relative observer is $f_r = f_m \sqrt{2c - v}/\sqrt{2c + v}$ from equation (4-38).

For the light source moving **away from** the receiver, the measurements of the frequency shift in electromagnetic waves can be combined by substituting equation (4-26) into equation (4-38) to obtain:

$$f_r = f_e \left(\frac{2c - v}{2c + v}\right) \tag{4-40}$$

where f_r is the frequency of the light pulse as measured by the relative observer, f_e is the frequency of the light pulse as measured by the event observer, v is the relative velocity between the event reference frame and the relative reference frame, and c is the speed of light.

The value for $(2c - v)/(2c + v)$ in equation (4-40) is always less than or equal to one, since the relative velocity is always less than **twice** the speed of light, due to the midpoint relative-velocity constraint. The relative observer measures a lower frequency, and a longer wavelength, when the **event** reference frame moves **away from** the **relative** reference frame, which is referred to as the **midpoint red-shift Doppler effect**.

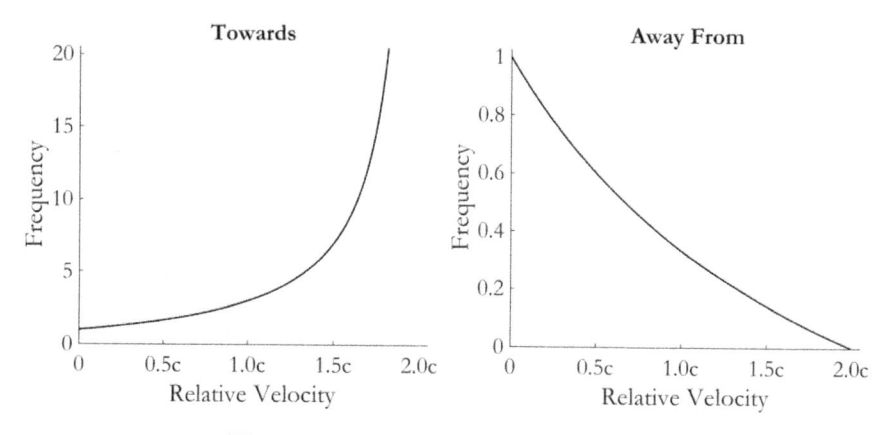

Figure 4-12. Midpoint Doppler effect

Figure 4-12 displays the frequency-shift values as measured by the relative observer against the relative-velocity values. To aid interpretation, all frequency-shift values are measured with a frequency of one ($f_e = 1$) in the event reference frame against different relative-velocity values, which are fractions of the speed of light. Table 4-5 reports a sample of the frequency-shift values against the relative-velocity values.

Table 4-5. Midpoint Doppler effect

Direction	Relative Velocity							
	0.00c	0.40c	0.80c	1.20c	1.60c	1.80c	1.90c	1.98c
Towards	1.00	1.50	2.33	4.00	9.00	19.00	39.00	199.00
Away From	1.00	0.67	0.43	0.25	0.11	0.05	0.03	0.01

The closer the relative velocity is to **twice** the speed of light, the larger the frequency shift in electromagnetic waves measured by the relative observer. For example, when the relative velocity is 1.98c with the light source moving **towards** the receiver, a frequency of one ($f_e = 1$) in the event reference frame is measured as a frequency of 199.00 by the relative observer. In contrast, when the relative velocity is 1.98c with the light source moving **away from** the receiver, a frequency of one ($f_e = 1$) in the event reference frame is measured as a frequency of 0.01 by the relative observer.

4.5 Relativity Comparison

This section compares the formula for measuring the frequency shift in electromagnetic waves for both special relativity and midpoint relativity. Table 4-6 reports the formula for the relative observer measuring the frequency shift in electromagnetic waves for a light source in the event reference frame.

Table 4-6. Frequency shift formula comparison

Direction	Special Relativity	Midpoint Relativity
Towards	$f_r = f_e \dfrac{\sqrt{c+v}}{\sqrt{c-v}}$	$f_r = f_e \left(\dfrac{2c+v}{2c-v}\right)$
Away From	$f_r = f_e \dfrac{\sqrt{c-v}}{\sqrt{c+v}}$	$f_r = f_e \left(\dfrac{2c-v}{2c+v}\right)$

The value of the relative velocity between the event reference frame and the relative reference frame drives the magnitude of the frequency shift in electromagnetic waves. In the midpoint theory of relativity, the frequency of an electromagnetic wave in the event reference frame is multiplied by either $(2c + v)/(2c - v)$ for the light source moving towards the receiver and $(2c - v)/(2c + v)$ for the light source moving away from the receiver. In contrast, in the special theory of relativity, the frequency of an electromagnetic wave in the event reference frame is multiplied by either $\sqrt{c+v}/\sqrt{c-v}$ for the light source moving towards the receiver and $\sqrt{c-v}/\sqrt{c+v}$ for the light source moving away from the receiver. These differences result in significantly different outcomes for both theories of relativity.

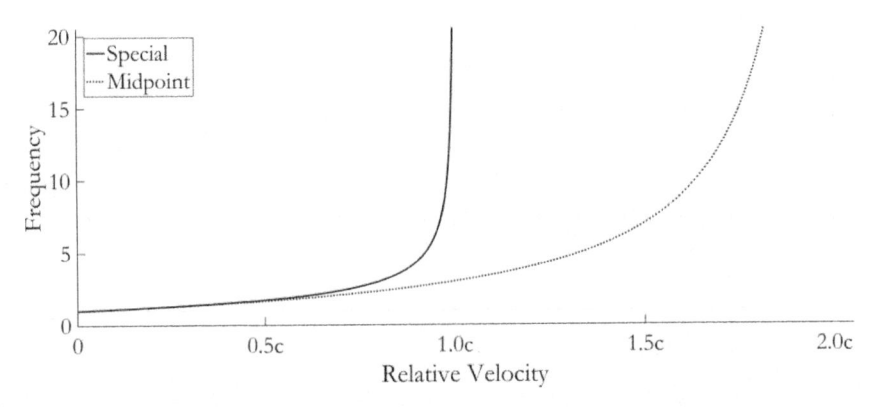

Figure 4-13. Comparison for the light source moving towards the receiver

The light source moving **towards** the receiver is compared first. Figure 4-13 displays the frequency-shift values as measured by the relative observer against the relative-velocity values, for both the special theory of relativity and the midpoint theory of relativity. To aid interpretation, all frequency-shift values are measured with a frequency of one ($f_e = 1$) in the event reference frame against different relative-velocity values, which are fractions of the speed of light. Table 4-7 reports a sample of the frequency-shift values against the relative-velocity values for both special relativity (SR) and midpoint relativity (MR).

Table 4-7. Comparison for the light source moving towards the receiver

	f_e	Relative Velocity							
		0.00c	0.50c	0.90c	0.95c	0.99c	1.80c	1.90c	1.98c
SR	1.00	1.00	1.73	4.36	6.24	14.11			
MR	1.00	1.00	1.67	2.64	2.81	2.96	19.00	39.00	199.00

For both special relativity and midpoint relativity, the relative observer will measure a **blue-shift** (higher frequency and shorter wavelength) when the light source moves towards the receiver. However, the special theory of relativity overestimates the strength of the frequency-shift compared to the midpoint theory of relativity. For example, when the relative velocity between the event reference frame and the relative reference frame is 0.99c, a frequency of one ($f_e = 1$) in the event reference frame is measured by the relative observer as a frequency of 14.11 for special relativity and a frequency of 2.96 for midpoint relativity.

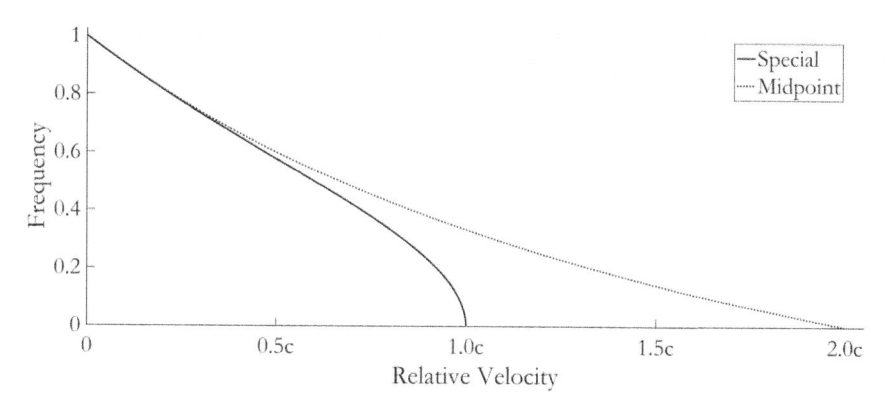

Figure 4-14. Comparison for the light source moving away from the receiver

The light source moving **away from** the receiver is now compared. Figure 4-14 displays the frequency-shift values as measured by the relative observer against the relative-velocity values, for both the special theory of relativity and the midpoint theory of relativity. To aid interpretation, all frequency-shift values are measured with a frequency of one ($f_e = 1$) in the event reference frame against different relative-velocity values, which are fractions of the speed of light. Table 4-8 reports a sample of the frequency-shift values against the relative-velocity values for both special relativity (SR) and midpoint relativity (MR).

Table 4-8. Comparison for the light source moving away from the receiver

	f_e	Relative Velocity							
		0.00c	0.50c	0.90c	0.95c	0.99c	1.80c	1.90c	1.98c
SR	1.00	1.00	0.58	0.23	0.16	0.07			
MR	1.00	1.00	0.60	0.38	0.36	0.34	0.05	0.03	0.01

For both special relativity and midpoint relativity, the relative observer will measure a **red-shift** (lower frequency and longer wavelength) when the light source moves away from the receiver. However, the special theory of relativity overestimates the strength of the frequency-shift compared to the midpoint theory of relativity. For example, when the relative velocity between the event reference frame and the relative reference frame is 0.99c, a frequency of one ($f_e = 1$) in the event reference frame is measured by the relative observer as a frequency of 0.07 for special relativity and a frequency of 0.34 for midpoint relativity.

The special theory of relativity is unable to measure frequency shifts in another reference frame when the relative velocity is greater than, or equal to, the speed of light. In addition, when the relative velocity is less than the speed of light, the special theory of relativity overestimates the frequency shift compared to the midpoint theory of relativity, with most of the overestimation occurring when the relative velocity is close to the speed of light.

5. Coordinate Transformations

Coordinate transformations express the relationship between the coordinates of space and time in one reference frame in terms of the coordinates of space and time in another reference frame. In this chapter, events can occur in **both** reference frames, rather than exclusively in the event reference frame. Consequently, a different naming convention is used for the two standard reference frames and their observers: a **left reference frame** with a **left observer** and a **right reference frame** with a **right observer**. This is in contrast to the previous naming convention of an event reference frame with an event observer and a relative reference frame with a relative observer.

The uniform motion of the reference frames is assumed to be parallel to the x-axis, so that the other space dimensions of y and z are perpendicular to the direction of uniform motion. It is well known that space dimensions perpendicular to the direction of uniform motion are not affected by relativity (see Young and Freedman, 2000). The time dimensions for both reference frames are assumed to coincide at the origin, so that the other dimensions are measured with respect to the origin.

5.1 Special Relativity

This section derives the coordinate transformations for the special theory of relativity, where there is a left reference frame with a left observer and a right reference frame with a right observer. This results in **two sets** of coordinate transformations. The first set expresses the coordinate system in the left reference frame in terms of the coordinate system in the right reference frame. The second set expresses the coordinate system in the right reference frame in terms of the coordinate system in the left reference frame.

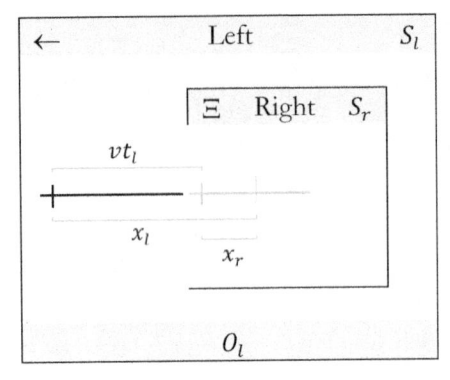

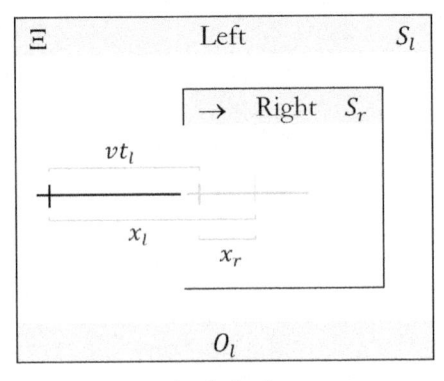

Figure 5-1. Two coordinates with respect to the left observer

With respect to the **left** observer, the special theory of relativity has two scenarios for the two reference frames. Figure 5-1 displays the x_l-space coordinate of the left reference frame S_l and the x_r-space coordinate of the right reference frame S_r, for both scenarios.

The two scenarios of special relativity for the left observer are:

1. The left reference frame S_l is in uniform motion to the left ← and the right reference frame S_r in a state of absolute rest Ξ.

2. The left reference frame S_l is in a state of absolute rest Ξ and the right reference frame S_r is in uniform motion to the right →.

The time dimensions of both reference frames are assumed to coincide at the origin. The two reference frames move apart a distance of vt_l in the time interval t_l. The special theory of relativity can be used by the left observer to measure the length of the x_r coordinate in the right reference frame. The x_l coordinate for the left observer can be written as:

$$x_l = vt_l + \frac{1}{\gamma_s} x_r \tag{5-1}$$

where x_r/γ_s is the relativistic distance of the x_r coordinate in the right reference frame, as measured by the left observer. Equation (5-1) can be rearranged in terms of the x_r coordinate to give:

$$\boxed{x_r = \gamma_s (x_l - vt_l)} \tag{5-2}$$

where x_r is the space coordinate in the right reference frame, γ_s is the special relativistic factor, x_l is the space coordinate in the left reference frame, v is the relative velocity between the left reference frame and the right reference frame, and t_l is the time coordinate in the left reference frame.

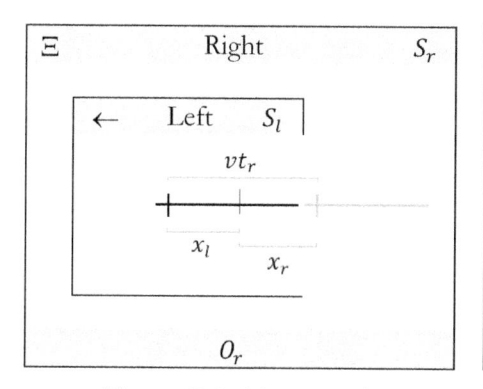

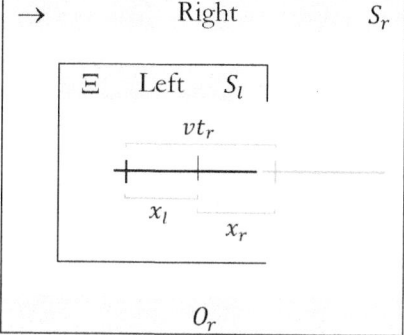

Figure 5-2. Two coordinates with respect to the right observer

With respect to the **right** observer, the special theory of relativity has two scenarios for the two reference frames. Figure 5-2 displays the x_l-space coordinate of the left reference frame S_l and the x_r-space coordinate of the right reference frame S_r, for both scenarios.

The two scenarios of special relativity for the right observer are:

1. The right reference frame S_r is in a state of absolute rest Ξ and the left reference frame S_l is in uniform motion to the left ←.

2. The right reference frame S_r is in uniform motion to the right → and the left reference frame S_l in a state of absolute rest Ξ.

The time dimensions of both reference frames are assumed to coincide at the origin. The two reference frames move apart a distance of vt_r in the time interval t_r. The special theory of relativity can be used by the right observer to measure the length of the x_l coordinate in the left reference frame. The x_r coordinate for the right observer can be written as:

$$x_r = -vt_r + \frac{1}{\gamma_s} x_l \qquad (5\text{-}3)$$

where x_l/γ_s is the relativistic distance of the x_l coordinate in the left reference frame, as measured by the right observer. Equation (5-3) can be rearranged in terms of the x_l coordinate to give:

$$\boxed{x_l = \gamma_s(x_r + vt_r)} \qquad (5\text{-}4)$$

where x_l is the space coordinate in the left reference frame, γ_s is the special relativistic factor, x_r is the space coordinate in the right reference frame, v is the relative velocity between the left reference frame and the right reference frame, and t_r is the time coordinate in the right reference frame.

The x-space coordinate transformations can be used to derive the t-time coordinate transformations. More specifically, equation (5-2) and equation (5-3) can be combined to eliminate the x_r coordinate:

$$\gamma_s(x_l - vt_l) = -vt_r + \frac{1}{\gamma_s}x_l \qquad (5\text{-}5)$$

Equation (5-5) can be divided by γ_s, and rearranged to give:

$$\frac{vt_r}{\gamma_s} = vt_l + \left(1 - \frac{v^2}{c^2}\right)x_l - x_l \qquad (5\text{-}6)$$

where $1/\gamma_s^2 = 1 - v^2/c^2$ from equation (2-6). Both sides of equation (5-6) can be multiplied by γ_s/v to solve for the t_r coordinate to obtain:

$$\boxed{t_r = \gamma_s\left(t_l - \frac{v}{c^2}x_l\right)} \qquad (5\text{-}7)$$

where t_r is the time coordinate in the right reference frame, γ_s is the special relativistic factor, t_l is the time coordinate in the left reference frame, v is the relative velocity between the left reference frame and the right reference frame, x_l is the space coordinate in the left reference frame, and c is the speed of light.

Similarly, equation (5-1) and equation (5-4) can be combined to eliminate the x_l coordinate:

$$\gamma_s(x_r + vt_r) = vt_l + \frac{1}{\gamma_s}x_r \qquad (5\text{-}8)$$

Equation (5-8) can be divided by γ_s, and rearranged to give:

$$\frac{vt_l}{\gamma_s} = vt_r + x_r - \left(1 - \frac{v^2}{c^2}\right)x_r \tag{5-9}$$

where $1/\gamma_s^2 = 1 - v^2/c^2$ from equation (2-6). Both sides of equation (5-9) can be multiplied by γ_s/v to solve for the t_l coordinate to obtain:

$$\boxed{t_l = \gamma_s\left(t_r + \frac{v}{c^2}x_r\right)} \tag{5-10}$$

where t_l is the time coordinate in the left reference frame, γ_s is the special relativistic factor, t_r is the time coordinate in the right reference frame, v is the relative velocity between the left reference frame and the right reference frame, x_r is the space coordinate in the right reference frame, and c is the speed of light.

Table 5-1. The special coordinate transformations

Dimension	Left	Right
t-time	$t_l = \gamma_s\left(t_r + \frac{v}{c^2}x_r\right)$	$t_r = \gamma_s\left(t_l - \frac{v}{c^2}x_l\right)$
x-space	$x_l = \gamma_s(x_r + vt_r)$	$x_r = \gamma_s(x_l - vt_l)$
y-space	$y_l = y_r$	$y_r = y_l$
z-space	$z_l = z_r$	$z_r = z_l$

Table 5-1 reports the two sets of **special coordinate transformations** for the special theory of relativity, which are known as **Lorentz transformations**. The first set of coordinate transformations express the coordinate system in the left reference frame in terms of the coordinate system in the right reference frame (see Left column). The second set of coordinate transformations express the coordinate system in the right reference frame in terms of the coordinate system in the left reference frame (see Right column). The uniform motion of the two reference frames is assumed to be parallel to the x_l and x_r axes, so that the other space dimensions of y and z are perpendicular to the direction of motion and are subsequently not affected.

5.2 Midpoint Relativity

This section derives the coordinate transformations for the midpoint theory of relativity. Midpoint relativity does **not** assume that the left reference frame or the right reference frame is in a state of absolute rest. In a system that consists of two **standard** reference frames in uniform motion, the midpoint is the only point that is in a state of absolute rest. Midpoint relativity creates a **phantom** midpoint reference frame with a phantom midpoint observer. The midpoint reference frame is always in a state of absolute rest.

The creation of a phantom midpoint reference frame divides the relativity environment into two parts: a **left-to-midpoint part** and a **midpoint-to-right part**. The coordinate transformations used for the special theory of relativity can be applied separately to both parts by correctly assuming that the midpoint reference frame is in a state of absolute rest. The coordinate transformations for each part can be combined to derive the coordinate transformations for the midpoint theory of relativity.

5.2.1 The left-to-midpoint part

This section derives the coordinate transformations for the left-to-midpoint part of the midpoint theory of relativity, where there is a left reference frame and a midpoint reference frame. This results in **two sets** of coordinate transformations. The first set expresses the coordinate system in the left reference frame in terms of the coordinate system in the midpoint reference frame. The second set expresses the coordinate system in the midpoint reference frame in terms of the coordinate system in the left reference frame.

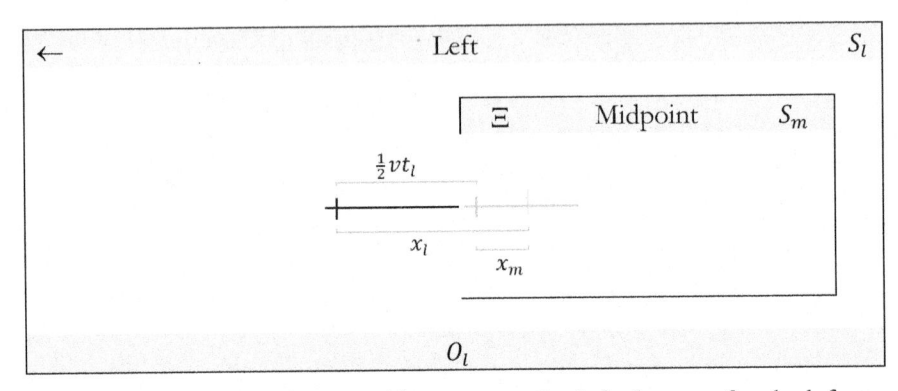

Figure 5-3. Two coordinates with respect to the left observer for the left-to-midpoint part

With respect to the **left** observer, the midpoint reference frame is assumed to be in a state of absolute rest and the left reference frame is in uniform motion to the left of the midpoint reference frame. Figure 5-3 displays the x_l coordinate in the left reference frame S_l and the x_m coordinate in the midpoint reference frame S_m.

The time dimensions of both reference frames are assumed to coincide at the origin. The left reference frame travels a distance of $\frac{1}{2}vt_l$ in the time interval t_l with respect to the midpoint reference frame. The length of the x_m coordinate in the midpoint reference frame is measured by the left observer using the event-to-midpoint part of the midpoint theory of relativity in equation (3-20). The x_l coordinate for the left observer can be written as:

$$x_l = \frac{v}{2}t_l + \frac{1}{\gamma_m}x_m \quad (5\text{-}11)$$

where x_m/γ_m is the relativistic distance of the x_m coordinate in the midpoint reference frame, as measured by the left observer. Equation (5-11) can be rearranged in terms of the x_m coordinate to obtain:

$$\boxed{x_m = \gamma_m\left(x_l - \frac{v}{2}t_l\right)} \quad (5\text{-}12)$$

where x_m is the space coordinate in the midpoint reference frame, γ_m is the midpoint relativistic factor, x_l is the space coordinate in the left reference frame, v is the relative velocity between the left reference frame and the right reference frame, and t_l is the time coordinate in the left reference frame.

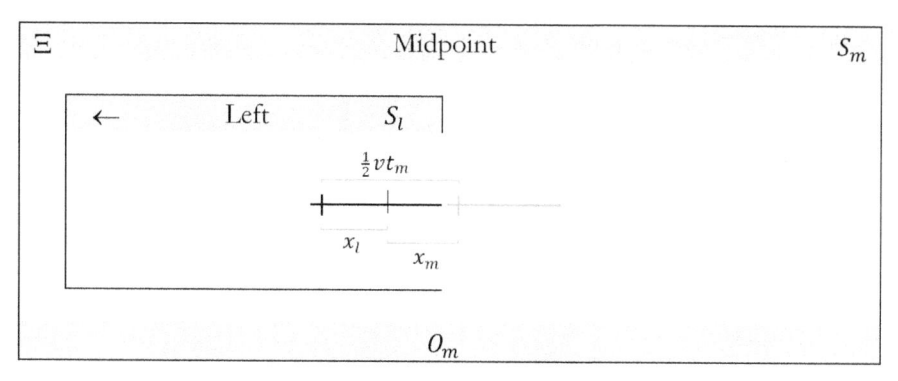

Figure 5-4. Two coordinates with respect to the midpoint observer for the left-to-midpoint part

With respect to the **midpoint** observer, the midpoint reference frame is assumed to be in a state of absolute rest and the left reference frame is in uniform motion to the left of the midpoint reference frame. Figure 5-4 displays the x_l coordinate of the left reference frame S_l and the x_m coordinate of the midpoint reference frame S_m.

The time dimensions of both reference frames are assumed to coincide at the origin. The left reference frame travels a distance of $\frac{1}{2}vt_m$ in the time interval t_m with respect to the midpoint reference frame. The length of the x_l coordinate in the left reference frame is measured by the midpoint observer using the event-to-midpoint part of the midpoint theory of relativity. The x_m coordinate for the midpoint observer can be written as:

$$x_m = -\frac{v}{2}t_m + \frac{1}{\gamma_m}x_l \tag{5-13}$$

where x_l/γ_m is the relativistic distance of the x_l coordinate in the left reference frame, as measured by the midpoint observer. Equation (5-13) can be rearranged in terms of the x_l coordinate to give:

$$\boxed{x_l = \gamma_m\left(x_m + \frac{v}{2}t_m\right)} \tag{5-14}$$

where x_l is the space coordinate in the left reference frame, γ_m is the midpoint relativistic factor, x_m is the space coordinate in the midpoint reference frame, v is the relative velocity between the left reference frame and the right reference frame, and t_m is the time coordinate in the midpoint reference frame.

The x-space coordinate transformations can be used to derive the t-time coordinate transformations. Equation (5-12) and equation (5-13) can be combined to eliminate the x_m coordinate and solved for the t_m coordinate to obtain:

$$\boxed{t_m = \gamma_m\left(t_l - \frac{v}{2c^2}x_l\right)} \tag{5-15}$$

where t_m is the time coordinate in the midpoint reference frame, γ_m is the midpoint relativistic factor, t_l is the time coordinate in the left reference frame, v is the relative velocity between the left reference frame and the right reference frame, x_l is the space coordinate in the left reference frame, and c is the speed of light.

Similarly, equation (5-11) and equation (5-14) can be combined to eliminate the x_l coordinate and solved for the t_l coordinate to obtain:

$$t_l = \gamma_m \left(t_m + \frac{v}{2c^2} x_m \right) \qquad (5\text{-}16)$$

where t_l is the time coordinate in the left reference frame, γ_m is the midpoint relativistic factor, t_m is the time coordinate in the midpoint reference frame, v is the relative velocity between the left reference frame and the right reference frame, x_m is the space coordinate in the midpoint reference frame, and c is the speed of light.

Table 5-2. The left-to-midpoint coordinate transformations

Dimension	Left	Midpoint
t-time	$t_l = \gamma_m \left(t_m + \frac{v}{2c^2} x_m \right)$	$t_m = \gamma_m \left(t_l - \frac{v}{2c^2} x_l \right)$
x-space	$x_l = \gamma_m \left(x_m + \frac{v}{2} t_m \right)$	$x_m = \gamma_m \left(x_l - \frac{v}{2} t_l \right)$
y-space	$y_l = y_m$	$y_m = y_l$
z-space	$z_l = z_m$	$z_m = z_l$

Table 5-2 reports the two sets of **left-to-midpoint coordinate transformations** for the midpoint theory of relativity. The first set of coordinate transformations express the coordinate system in the left reference frame in terms of the coordinate system in the midpoint reference frame (see Left column). The second set of coordinate transformations express the coordinate system in the midpoint reference frame in terms of the coordinate system in the left reference frame (see Midpoint column). The uniform motion of the two reference frames is assumed to be parallel to the x_l and x_m axes, so that the other space dimensions of y and z are perpendicular to the direction of motion and are subsequently not affected.

5.2.2 The midpoint-to-right part

This section derives the coordinate transformations for the midpoint-to-right part of the midpoint theory of relativity, where there is a midpoint reference frame and a right reference frame. This results in **two sets** of coordinate transformations. The first set expresses the coordinate system in the midpoint reference frame in terms of the coordinate system in the right reference frame. The second set expresses the coordinate system in the right reference frame in terms of the coordinate system in the midpoint reference frame.

With respect to the **midpoint** observer, the midpoint reference frame is assumed to be in a state of absolute rest and the right reference frame is in uniform motion to the right of the midpoint reference frame. Figure 5-5 displays the x_m coordinate of the midpoint reference frame S_m and the x_r coordinate of the right reference frame S_r.

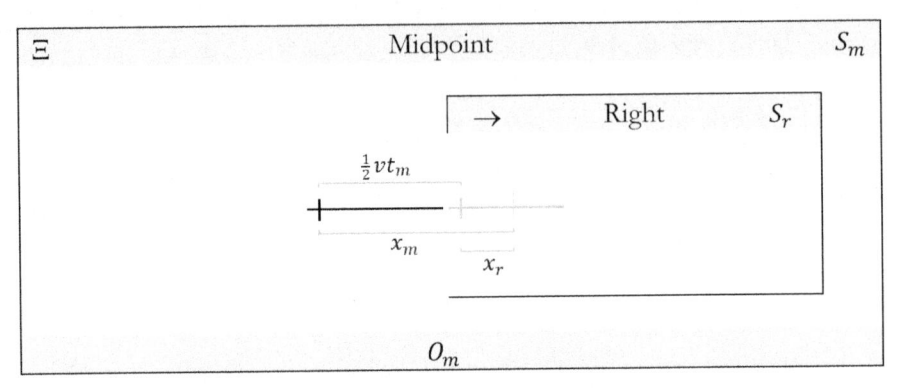

Figure 5-5. Two coordinates with respect to the midpoint observer for the midpoint-to-right part

The time dimensions of both reference frames are assumed to coincide at the origin. The right reference frame travels a distance of $\frac{1}{2}vt_m$ in the time interval t_m with respect to the midpoint reference frame. The midpoint-to-relative part of the midpoint theory of relativity can be used by the midpoint observer to measure the length of the x_r coordinate in the right reference frame. The x_m coordinate for the midpoint observer can be written as:

$$x_m = \frac{v}{2}t_m + \frac{1}{\gamma_m}x_r \qquad (5\text{-}17)$$

where x_r/γ_m is the relativistic distance of the x_r coordinate in the right reference frame, as measured by the midpoint observer. Equation (5-17) can be rearranged in terms of the x_r coordinate to produce:

$$\boxed{x_r = \gamma_m\left(x_m - \frac{v}{2}t_m\right)} \qquad (5\text{-}18)$$

where x_r is the space coordinate in the right reference frame, γ_m is the midpoint relativistic factor, x_m is the space coordinate in the midpoint reference frame, v is the relative velocity between the left reference frame and the right reference frame, and t_m is the time coordinate in the midpoint reference frame.

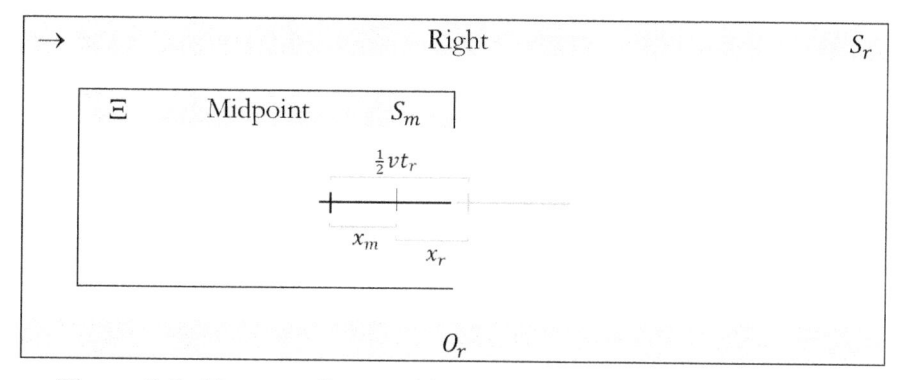

Figure 5-6. Two coordinates with respect to the right observer for the midpoint-to-right part

With respect to the **right** observer, the midpoint reference frame is assumed to be in a state of absolute rest and the right reference frame is in uniform motion to the right of the midpoint reference frame. Figure 5-6 displays the x_m coordinate of the midpoint reference frame S_m and the x_r coordinate of the right reference frame S_r.

The time dimensions of both reference frames are assumed to coincide at the origin. The right reference frame travels a distance of $\frac{1}{2}vt_r$ in the time interval t_r with respect to the midpoint reference frame. The midpoint-to-relative part of the midpoint theory of relativity can be used by the right observer to measure the length of the x_m coordinate in the midpoint reference frame. The x_r coordinate for the right observer can be written as:

$$x_r = -\frac{v}{2}t_r + \frac{1}{\gamma_m}x_m \tag{5-19}$$

where x_m/γ_m is the relativistic distance of the x_m coordinate in the midpoint reference frame, as measured by the right observer. Equation (5-19) can be rearranged in terms of the x_m coordinate to produce:

$$\boxed{x_m = \gamma_m\left(x_r + \frac{v}{2}t_r\right)} \tag{5-20}$$

where x_m is the space coordinate in the midpoint reference frame, γ_m is the midpoint relativistic factor, x_r is the space coordinate in the right reference frame, v is the relative velocity between the left reference frame and the right reference frame, and t_r is the time coordinate in the right reference frame.

The x-space coordinate transformations can be used to derive the t-time coordinate transformations. Equation (5-18) and equation (5-19) can be combined to eliminate the x_r coordinate and solved for the t_r coordinate:

$$t_r = \gamma_m \left(t_m - \frac{v}{2c^2} x_m \right) \qquad (5\text{-}21)$$

where t_r is the time coordinate in the right reference frame, γ_m is the midpoint relativistic factor, t_m is the time coordinate in the midpoint reference frame, v is the relative velocity between the left reference frame and the right reference frame, x_m is the space coordinate in the midpoint reference frame, and c is the speed of light.

Similarly, equation (5-17) and equation (5-20) can be combined to eliminate the x_m coordinate and solved for the t_m coordinate:

$$t_m = \gamma_m \left(t_r + \frac{v}{2c^2} x_r \right) \qquad (5\text{-}22)$$

where t_m is the time coordinate in the midpoint reference frame, γ_m is the midpoint relativistic factor, t_r is the time coordinate in the right reference frame, v is the relative velocity between the left reference frame and the right reference frame, x_r is the space coordinate in the right reference frame, and c is the speed of light.

Table 5-3. The midpoint-to-right coordinate transformations

Dimension	Midpoint	Right
t-time	$t_m = \gamma_m \left(t_r + \frac{v}{2c^2} x_r \right)$	$t_r = \gamma_m \left(t_m - \frac{v}{2c^2} x_m \right)$
x-space	$x_m = \gamma_m \left(x_r + \frac{v}{2} t_r \right)$	$x_r = \gamma_m \left(x_m - \frac{v}{2} t_m \right)$
y-space	$y_m = y_r$	$y_r = y_m$
z-space	$z_m = z_r$	$z_r = z_m$

Table 5-3 reports the two sets of **midpoint-to-right coordinate transformations** for the midpoint theory of relativity. The first set of coordinate transformations express the coordinate system in the midpoint reference frame in terms of the coordinate system in the right reference frame (see Midpoint column). The second set of coordinate transformations express the coordinate system in the right reference frame in terms of the coordinate system in the midpoint reference frame (see Right column). The uniform motion of the two reference frames is assumed to be parallel to the x_m and x_r axes, so that the other space dimensions of y and z are perpendicular to the direction of motion and are subsequently not affected.

5.2.3 The complete solution

The midpoint theory of relativity divides the relativity environment into two parts: a left-to-midpoint part and a midpoint-to-right part. The previous sections derived coordinate transformations for each part. This section combines the coordinate transformations for both parts to derive the **midpoint coordinate transformations** for the midpoint theory of relativity. This results in **two sets** of coordinate transformations. The first set expresses the coordinate system in the left reference frame in terms of the coordinate system in the right reference frame. The second set expresses the coordinate system in the right reference frame in terms of the coordinate system in the left reference frame.

The **first set** of coordinate transformations for both parts of midpoint relativity can be combined. The coordinate transformation for the t_l coordinate can be derived by substituting the x_m coordinate from equation (5-20) and the t_m coordinate from equation (5-22) into equation (5-16) to obtain:

$$\begin{aligned} t_l &= \gamma_m \left(t_m + \frac{v}{2c^2} x_m \right) \\ &= \gamma_m^2 \left(t_r + \frac{v}{c^2} x_r + \frac{v^2}{4c^2} t_r \right) \\ &= \gamma_m^2 \left(\delta_m t_r + \frac{v}{c^2} x_r \right) \end{aligned} \qquad (5\text{-}23)$$

where t_l is the time coordinate in the left reference frame, γ_m is the midpoint relativistic factor, t_r is the time coordinate in the right reference frame, v is the relative velocity between the left reference frame and the right reference frame, x_r is the space coordinate in the right reference frame, c is the speed of light, and δ_m is the **midpoint scale factor** defined by:

$$\boxed{\delta_m = 1 + \frac{v^2}{4c^2}} \qquad (5\text{-}24)$$

Similarly, the coordinate transformation for the x_l coordinate can be derived by substituting the x_m coordinate from equation (5-20) and the t_m coordinate from equation (5-22) into equation (5-14) to obtain:

$$\boxed{\begin{aligned} x_l &= \gamma_m \left(x_m + \frac{v}{2} t_m \right) \\ &= \gamma_m^2 \left(x_r + v t_r + \frac{v^2}{4c^2} x_r \right) \\ &= \gamma_m^2 (\delta_m x_r + v t_r) \end{aligned}} \qquad (5\text{-}25)$$

where x_l is the space coordinate in the left reference frame, γ_m is the midpoint relativistic factor, δ_m is the midpoint scale factor, x_r is the space coordinate in the right reference frame, t_r is the time coordinate in the right reference frame, and v is the relative velocity between the left reference frame and the right reference frame.

The **second set** of coordinate transformations for both parts of midpoint relativity can be combined. The coordinate transformation for the t_r coordinate can be derived by substituting the x_m coordinate from equation (5-12) and the t_m coordinate from equation (5-15) into equation (5-21) to obtain:

$$\boxed{\begin{aligned} t_r &= \gamma_m \left(t_m - \frac{v}{2c^2} x_m \right) \\ &= \gamma_m^2 \left(t_l - \frac{v}{c^2} x_l + \frac{v^2}{4c^2} t_l \right) \\ &= \gamma_m^2 \left(\delta_m t_l - \frac{v}{c^2} x_l \right) \end{aligned}} \qquad (5\text{-}26)$$

where t_r is the time coordinate in the right reference frame, γ_m is the midpoint relativistic factor, δ_m is a midpoint scale factor, t_l is the time coordinate in the left reference frame, v is the relative velocity between the left reference frame and the right reference frame, x_l is the space coordinate in the left reference frame, and c is the speed of light.

Similarly, the coordinate transformation for the x_r coordinate can be derived by substituting the x_m coordinate from equation (5-12) and the t_m coordinate from equation (5-15) into equation (5-18) to obtain:

$$\begin{aligned} x_r &= \gamma_m \left(x_m - \frac{v}{2} t_m \right) \\ &= \gamma_m^2 \left(x_l - v t_l + \frac{v^2}{4c^2} x_l \right) \\ &= \gamma_m^2 (\delta_m x_l - v t_l) \end{aligned} \quad (5\text{-}27)$$

where x_r is the space coordinate in the right reference frame, γ_m is the midpoint relativistic factor, δ_m is a midpoint scale factor, x_l is the space coordinate in the left reference frame, t_l is the time coordinate in the left reference frame, and v is the relative velocity between the left reference frame and the right reference frame.

Table 5-4. The midpoint coordinate transformations

Dimension	Left	Right
t-time	$t_l = \gamma_m^2 \left(\delta_m t_r + \frac{v}{c^2} x_r \right)$	$t_r = \gamma_m^2 \left(\delta_m t_l - \frac{v}{c^2} x_l \right)$
x-space	$x_l = \gamma_m^2 (\delta_m x_r + v t_r)$	$x_r = \gamma_m^2 (\delta_m x_l - v t_l)$
y-space	$y_l = y_r$	$y_r = y_l$
z-space	$z_l = z_r$	$z_r = z_l$

Table 5-4 reports the two sets of **midpoint coordinate transformations** for the midpoint theory of relativity. The first set of coordinate transformations express the coordinate system in the left reference frame in terms of the coordinate system in the right reference frame (see Left column). The second set of coordinate transformations express the coordinate system in the right reference frame in terms of the coordinate system in the left reference frame (see Right column). The uniform motion of the two reference frames is assumed to be parallel to the x_l and x_r axes, so that the other space dimensions of y and z are perpendicular to the direction of motion and are subsequently not affected.

5.3 Relativity Comparison

This section compares the coordinate transformations for both the special theory relativity and the midpoint theory of relativity. Table 5-5 reports the coordinate transformations for the coordinate system in the **left** reference frame in terms of the coordinate system in the right reference frame for both special relativity and midpoint relativity.

Table 5-5. Coordinate transformations comparison for the left reference frame

Dimension	Special Relativity	Midpoint Relativity
t-time	$t_l = \gamma_s\left(t_r + \dfrac{v}{c^2}x_r\right)$	$t_l = \gamma_m^2\left(\delta_m t_r + \dfrac{v}{c^2}x_r\right)$
x-space	$x_l = \gamma_s(x_r + vt_r)$	$x_l = \gamma_m^2(\delta_m x_r + vt_r)$
y-space	$y_l = y_r$	$y_l = y_r$
z-space	$z_l = z_r$	$z_l = z_r$

Similarly, Table 5-6 reports the coordinate transformations for the coordinate system in the **right** reference frame in terms of the coordinate system in the left reference frame for both special relativity and midpoint relativity.

Table 5-6. Coordinate transformations comparison for the right reference frame

Dimension	Special Relativity	Midpoint Relativity
t-time	$t_r = \gamma_s\left(t_l - \dfrac{v}{c^2}x_l\right)$	$t_r = \gamma_m^2\left(\delta_m t_l - \dfrac{v}{c^2}x_l\right)$
x-space	$x_r = \gamma_s(x_l - vt_l)$	$x_r = \gamma_m^2(\delta_m x_l - vt_l)$
y-space	$y_r = y_l$	$y_r = y_l$
z-space	$z_r = z_l$	$z_r = z_l$

For the special theory of relativity, coordinate transformations are not possible when the relative velocity is greater than, or equal to, the speed of light. In contrast, for the midpoint theory of relativity, coordinate transformations are possible when relative velocity is less than **twice** the speed of light.

6. Conclusion

This book addresses two inherent problems with the special theory of relativity. The two problems are the **at-rest assumption** and the **relative-velocity constraint**. The at-rest assumption requires that one of the two standard reference frames is in a state of absolute rest. The relative-velocity constraint enforces that the relative velocity between the two standard reference frames is less than the speed of light.

The special theory of relativity assumes that one of the two standard reference frames is in a state of absolute rest. However, there is no way of knowing which one of the standard reference frames is in a state of absolute rest. The question then arises: which standard reference frame does one choose? This book has shown that the answer is **neither**.

The midpoint theory of relativity does not assume that either of the two standard reference frames is in a state of absolute rest. In a system that consists of two standard reference frames in uniform motion, the midpoint is the only point that is always in a state of absolute rest. The midpoint theory of relativity creates a **phantom midpoint reference frame** with a **phantom midpoint observer** at the midpoint. The midpoint reference frame is always in a state of absolute rest.

The creation of a midpoint reference frame divides the relativity environment into two parts: a one-standard-reference-frame-to-the-midpoint-reference-frame part and a midpoint-reference-frame-to-the-other-standard-reference-frame part. The special theory of relativity can be applied to both parts by correctly assuming that the midpoint reference frame is in a state of absolute rest. The solution for each part can be combined to derive a complete solution for the midpoint theory of relativity.

The special theory of relativity is unable to derive measurements when the relative velocity between the two standard reference frames is greater than the speed of light. In contrast, the midpoint theory of relativity is able to derive measurements when the relative velocity between the two standard reference frames is less than **twice** the speed of light. When the relative velocity is less than the speed of light, the special theory of relativity overestimates time dilation, length contraction, and the frequency shift in electromagnetic waves, compared to the midpoint theory of relativity.

Bibliography

Einstein, A., 1905. Zur Elektrodynamik bewegter Körper. Annalen der Physik 322, 10, 891–921.

Einstein, A., 1916. Relativity: The Special and General Theory. Princeton University Press.

Mirimanoff, D., 1921. La transformation de Lorentz-Einstein et le temps universel de M. Ed. Guillaume. Archives des sciences physiques et naturelles (Supplement). 5. 3: 46–48.

Young, H.D., and Freedman, R.A., 2000. University Physics, Addison-Wesley Series in Physics.

Index

Albert Einstein, 2

At-rest assumption
midpoint relativity, 6
special relativity, 2

Coordinate transformations
defined, 71
midpoint relativity, 76
relativity comparison, 86
special relativity, 71

Frequency shift
midpoint relativity, 54
relativity comparison, 68
special relativity, 49

Gamma
midpoint relativity, 17
special relativity, 14

Length contraction
defined, 1
midpoint relativity, 43
special relativity, 31

Length of objects
midpoint relativity, 32
relativity comparison, 44
special relativity, 27

Light-speed constraint, 2

Lorentz transformations
defined, 1, 75

Midpoint Doppler effect
blue-shift, 66
red-shift, 66

Midpoint relativistic factor, 8, 17

Midpoint relativity
coordinate transformations, 76
defined, 5
frequency shift, 54
length contraction, 43
length of objects, 32
reference frames, 6
the theory of, 5
time dilation, 22
time intervals, 16

Midpoint scale factor, 83

Observers
event, 2, 6
left, 71
midpoint, 5
relative, 2, 6
right, 71

Postulates
principle of relativity, 2

speed of light invariance, 2

Principle of relativity
postulate, 2

Reference frames
accelerated, 2
event, 2, 6
inertial, 2
left, 71
non-inertial, 2
relative, 2, 6
right, 71

Relative velocity
defined, 2

Relative-velocity constraint
midpoint relativity, 8
special relativity, 4

Relativistic Doppler effect
defined, 1

Relativistic factor
midpoint relativity, 8, 17
special relativity, 4, 14

Relativity comparison
coordinate transformations, 86
frequency shift, 68
length of objects, 44
time intervals, 23

Special Doppler effect
blue-shift, 51
red-shift, 53

Special relativistic factor, 4, 14

Special relativity
coordinate transformations, 71
defined, 2
frequency shift, 49
length contraction, 31
length of objects, 27
reference frames, 2
time dilation, 15
time intervals, 13

Speed of light invariance
postulate, 2

Time dilation
defined, 1
midpoint relativity, 22
special relativity, 15

Time intervals
midpoint relativity, 16
relativity comparison, 23
special relativity, 13

Printed in Great Britain
by Amazon